Knowledge Management and Innovation in Network Organizations:

Emerging Research and Opportunities

Jerzy Kisielnicki
Warsaw University, Poland

Olga Sobolewska
Warsaw University of Technology, Poland

A volume in the Advances in
Business Information Systems and
Analytics (ABISA) Book Series

Published in the United States of America by
    IGI Global
    Business Science Reference (an imprint of IGI Global)
    701 E. Chocolate Avenue
    Hershey PA, USA 17033
    Tel: 717-533-8845
    Fax: 717-533-8661
    E-mail: cust@igi-global.com
    Web site: http://www.igi-global.com

Library of Congress Cataloging-in-Publication Data

Names: Kisielnicki, Jerzy, author. | Sobolewska, Olga, 1980- author.
Title: Knowledge management and innovation in network organizations :
  emerging research and opportunities / by Jerzy Kisielnicki and Olga
  Sobolewska.
Description: Hershey, PA : Business Science Reference, [2019]
Identifiers: LCCN 2017058244| ISBN 9781522559306 (hardcover) | ISBN
  9781522559313 (ebook)
Subjects: LCSH: Knowledge management. | Information technology--Management. |
  Technological innovations--Management. | Virtual reality in management. |
  Telematics.
Classification: LCC HD30.2 .K5686 2019 | DDC 658.4/038--dc23 LC record available at https://
lccn.loc.gov/2017058244

This book is published in the IGI Global book series Advances in Business Information Systems and Analytics (ABISA) (ISSN: 2327-3275; eISSN: 2327-3283)

British Cataloguing in Publication Data
A Cataloguing in Publication record for this book is available from the British Library.

All work contributed to this book is new, previously-unpublished material.
The views expressed in this book are those of the authors, but not necessarily of the publisher.

For electronic access to this publication, please contact: eresources@igi-global.com.

Advances in Business Information Systems and Analytics (ABISA) Book Series

ISSN:2327-3275
EISSN:2327-3283

Editor-in-Chief: Madjid Tavana, La Salle University, USA

MISSION

The successful development and management of information systems and business analytics is crucial to the success of an organization. New technological developments and methods for data analysis have allowed organizations to not only improve their processes and allow for greater productivity, but have also provided businesses with a venue through which to cut costs, plan for the future, and maintain competitive advantage in the information age.

The **Advances in Business Information Systems and Analytics (ABISA) Book Series** aims to present diverse and timely research in the development, deployment, and management of business information systems and business analytics for continued organizational development and improved business value.

COVERAGE

- Data Governance
- Strategic Information Systems
- Data Strategy
- Legal information systems
- Information Logistics
- Business Systems Engineering
- Algorithms
- Data Analytics
- Business Intelligence
- Business Information Security

IGI Global is currently accepting manuscripts for publication within this series. To submit a proposal for a volume in this series, please contact our Acquisition Editors at Acquisitions@igi-global.com or visit: http://www.igi-global.com/publish/.

Titles in this Series

For a list of additional titles in this series, please visit:
https://www.igi-global.com/book-series/advances-business-information-systems-analytics/37155

Harnessing Human Capital Analytics for Competitive Avantage
Mohit Yadav (BML Munjal University, India) Shrawan Kumar Trivedi (Indian Institute of Management Sirmaur, India) Anil Kumar (BML Munjal University, India) and Santosh Rangnekar (Indian Institute of Technology Roorkee, India)
Business Science Reference • ©2018 • 367pp • H/C (ISBN: 9781522540380) • US $215.00

Corporate Social Responsibility for Valorization of Cultural Organizations
María del Pilar Muñoz Dueñas (University of Vigo, Spain) Lucia Aiello (Sapienza University of Rome, Italy) Rosario Cabrita (University Nova de Lisboa, Portugal) and Mauro Gatti (Sapienza University of Rome, Italy)
Business Science Reference • ©2018 • 328pp • H/C (ISBN: 9781522535515) • US $215.00

Contemporary Identity and Access Management Architectures Emerging Research ...
Alex Chi Keung Ng (Federation University, Australia)
Business Science Reference • ©2018 • 241pp • H/C (ISBN: 9781522548287) • US $175.00

Research, Practices, and Innovations in Global Risk and Contingency Management
Kenneth David Strang (State University of New York, USA & APPC Research, Australia) Maximiliano E. Korstanje (University of Palermo, Argentina) and Narasimha Vajjhala (American University of Nigeria, Nigeria)
Business Science Reference • ©2018 • 418pp • H/C (ISBN: 9781522547549) • US $245.00

Modeling and Simulation Techniques for Improved Business Processes
Maryam Ebrahimi (Azad University, Iran)
Business Science Reference • ©2018 • 246pp • H/C (ISBN: 9781522532262) • US $185.00

Value Sharing for Sustainable and Inclusive Development
Mario Risso (Niccolò Cusano University, Italy) and Silvia Testarmata (Niccolò Cusano University, Italy)
Business Science Reference • ©2018 • 398pp • H/C (ISBN: 9781522531470) • US $245.00

For an entire list of titles in this series, please visit:
https://www.igi-global.com/book-series/advances-business-information-systems-analytics/37155

701 East Chocolate Avenue, Hershey, PA 17033, USA
Tel: 717-533-8845 x100 • Fax: 717-533-8661
E-Mail: cust@igi-global.com • www.igi-global.com

Table of Contents

Preface

The formation of network organizations is underpinned by the development of ICTs (Information Communication Technologies). This monograph seeks to present and examine network organization management in a comprehensive manner. Special emphasis is put on the role of network organizations in the context of knowledge management. Such organizations promote the application of an innovation-oriented approach. The monograph depicts theoretical and practical issues arising from our research on the transformation of management structures of various organizations. The research shows that the application of a network management model allows for significant economic effects to be obtained both for the constituent organizations and for the society. Creating a network organization may improve its competitiveness and ensure its higher international ratings. The emergence of network organizations ensues from the evolution of traditional linear structures. The transformation process could take place thanks to substantial progress in ICT development.

We put forward the thesis that a network organization is an appropriate management infrastructure for knowledge-based organizations. Relations between the elements of a network organization are realized as various forms of links that are defined by adopted procedures, communication protocols and the applicable legislation. The latter is not always adapted to the changing world, often failing to keep up with the developments in modern technology. What is specifically characteristic of cyberspace is multi-directional connections and lack of possibility to establish its boundaries by physical means. It is assumed and justified that a network form is suited for open, adaptive and generative organizations. This is confirmed by the latest social and economic trends that indicate the need for cooperation and emphasis on the importance of communication systems.

Network organizations may be divided into two categories: those formed as a result of transformation of existing traditional organizations and "emerging" organizations that build a network of relations on a voluntary basis. The monograph portrays both types. We analyze the global network of a national and multinational organization, pointing to the issues of the interpenetration of cultures. The monograph highlights the role of organizations as network accelerators in creating a multicultural world – in business, science and everyday life of people today. It should be noted that multinational network organizations contribute to the transfer of knowledge from organizations in developed countries to those in developing countries, thereby reducing the latter's delays in innovation processes.

The presentation logic herein is as follows. The entire material is divided into six intertwined thematic parts. The contents can also be presented graphically as circles, where the core is represented by the first chapter discussing network structures as a development path for contemporary organizations. This particularly concerns the study of the role of organizations in: knowledge management, development of innovation and science, and establishment of transnational cooperation. The second chapter explores innovation issues. Its first and second sections examine the determinants of knowledge flow and network organization development. This is the starting point for an empirical description of network organizations as carriers of knowledge and innovation. The third chapter analyzes technical resources in such organizations. It focuses on the Internet as a critical infrastructure of network organizations and investigates an impact of ICT development on the operation of organizations (Big Data, Cloud computing, BI systems). The fourth chapter considers what connects the individual links of an organization – it analyzes communication processes within traditional and network organizations. Both communication determinants and tools in network organizations, including loss of knowledge, are examined in greater detail. That part is an "overlay" on previous chapters and is very closely related to them. In modern management and applications, it is difficult to distinguish management with no communication. A comparative analysis of traditional and network communication systems allows for presenting the advantage that network organizations have over traditional ones. The fifth chapter contains an examination of decision-making processes. What we have achieved, what we do, what we strive for, what we are and what we want to be is the result of our ongoing decisions. The development of society, its culture, wealth, quality of existence and well-being are the product of many yesterday's and today's choices. Decisions are made and implemented by people of varying professional backgrounds, attached to different value systems, and pursuing different goals. In that chapter, decision-making processes are analyzed in two types of organizations: traditional and network ones. The monograph concludes with the sixth chapter. It puts forward the thesis that the implementation of an ICT-supported organizational development strategy transforms local organizations into global ones. Organizational structures, forming part of the infrastructure for the implemented strategy, change. As a consequence, the transition from hierarchical structures to network structures takes place. The transformation results in the rise of global network organizations.

The research on the functioning of network organizations makes it possible to argue that the network system has an advantage over traditional organizations, notably as regards the following elements:

1. Monitoring. Risks to implementation and deviations from planned costs, time, adopted parameters are spotted earlier than in hierarchical organizations. Staff working in teams that operate within network systems feel responsible for the organization;
2. Knowledge transfer in task performance. There is good cooperation between workers in a network as regards knowledge transfer and innovation policy implementation. No artificial barriers exist such as leaders and employees.
3. Problem solving. Conflicts in task implementation are much less severe than in traditional organizations, and if they arise, they are quickly resolved within the project team.

Whenever a monograph is written, a question should be asked about the target reader. Therefore, the following questions should be answered:

1. Who are the readers and what theoretical knowledge and experience do they have?
2. What are their intellectual needs and what knowledge do they wish to gain?

Finally, there is the concern whether the study will be so attractive that, with such a wealth of management and information technology publications, someone would like to spend some time reading it. This is not coquetry. Authors always believe that their work is interesting and useful. Without this conviction, our efforts would be doomed to immediate failure. We seem to have a professional and scientific background to write a monograph anchored both in management and in MIS.

We envisage the readership profile as follows:

1. Researchers who wish to deepen their knowledge and confront their judgments,
2. Students of economic or technical universities who are seeking support in studying subjects such as: management, information technologies, introduction to business informatics, design of IT and information systems for management,
3. Heads of various management levels who wish to raise their awareness about knowledge management, the operation of network organizations, and who have completed their university studies and are currently seeking to update their knowledge or have graduated from a faculty unrelated to management,
4. Executive employees of various organizations who would like to find out about contemporary management systems for various reasons – to get promoted, for example,
5. People who wish to learn about contemporary management issues and principles of building an information society and who cooperate with IT specialists in project implementation so that they can establish "intellectual contact" with such specialists.

The monograph is universal. It is a compendium of knowledge about network organizations, their structures and management systems. The analyses presented herein include our own research and practical managerial experience. The monograph can be recommended in academic curricula comprising scientific organizational and management theories, in particular strategic management, management information systems and project management. It will also provide support for post-graduate and MBA students.

We would like to thank Mahdi Khosrow-Pour and Eli Cohen, who allowed us to verify certain theses in an international environment. We are also grateful to the reviewers and consultants who have contributed to the individual parts of the book. The list of acknowledgments is incomplete, yet we would also like to thank post-graduate students and participants of a doctoral seminar at the Faculties of Management of the University of Warsaw and Warsaw University of Technology for discussion and participation in our research.

Chapter 1

The Network Organization as the Development Path for Modern Organizations

ABSTRACT

The first chapter addresses the phenomenon and concept of a network organization. It is a form that is more and more present in today's digital world. It does not solely concern typical network, virtual organizations that are established and operate only on the internet. We ever more frequently see organizations that go beyond their own traditional organizational structures. They are often interdisciplinary and cover a wide range of seemingly different areas of life and economy. Network organizations are particularly common where knowledge and innovation are created; therefore, they are extremely popular when research and development (R&D) and scientific projects are implemented. How widespread network structures are in the life of today's organizations will be established by means of the survey results that will be described in this chapter and in the further part of this monograph.

NETWORK ORGANIZATION: THE CONCEPT AND ITS EVOLUTION, NETWORKS, AND NODES

The origins of the term "organization" can be found in Greek, where the word *organon* (Lat. *organum*) signified a specialized part separated from the environment in order to perform a strictly defined function. This meaning can be easily found in contemporary definitions of the organization, where the

DOI: 10.4018/978-1-5225-5930-6.ch001

main emphasis is on a clear structure and a common goal of action. This is evident in the approach proposed by T. Kotarbiński (1958,75), who writes that an organization is "a kind of entirety, given the relation of its own elements to it, namely the entirety whose all components contribute to its success." A similar wording is present in G. Hostelet's approach, where the entirety has been formed from cooperating parts. This collaboration is intentional, as can be seen in the definition put forward by R.W. Griffin, where an organization is "a group of people working together in structured and coordinated fashion to achieve a set of goals." (1996, 35). R. Ackoff specifies the concept of collaboration and points out that achieving the set goals requires division of labor, introduction of an information exchange system and a command and control system (Bielski, 2004, 35). Interestingly, these classical definitions treat organizations in a goal-based manner. The separation of a specific system, namely an organization, from the environment serves the achievement of common goals. In theory, organizations go beyond the ownership and formal structure. They take the form of networks of links and relationships present in the market (Figure 1).

Figure 1. Business relationships and networks – a focal firm perspective
Source: (Möller, Halinen, 1999, 415)

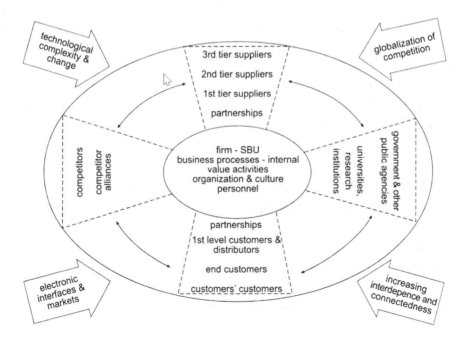

The digital revolution has covered management to a large extent (Brynjolfsson, McAfee, 2015). This influence is evident both in the take-over of certain areas of organization management by IT systems (e.g. production automation) and through changes in organizations themselves and their organizational structures. Classical organizations, structured in a hierarchical fashion, follow a unique path of evolution. What could be observed in the 1960s was the transformation of classical linear organizational relationships into matrix structures intended to streamline management and facilitate project management (Stoner & Wankel, 1996, 218-219). Nowadays, typical network organizations are ever more often emerging. The rise of such organizations is largely dependent on the development of networks and information and communication technologies (ICTs). The network-based nature of an organization does not only mean how the structure is organized formally but, to a much greater degree, how it operates. The importance of ICT infrastructure for the formation and operation of network organizations is described in a further section of this manuscript (Chapter 3). Network organizations are oftentimes established to make knowledge exchange better and to improve the efficiency of knowledge management.

They emerge as a result of the transformation of traditional structures organized in a hierarchical way into modern forms relying on the use of ICTs. These are structures that are not affected by functional and geographical constraints as much as classical organizations. Their driving force is an IT network allowing them to cross both geographic and institutional boundaries without much impediment. M. Castells even speaks of a new type of economy (information economy) that is replacing the old-fashioned industrial approach (2007, 189). In the previous type of economy, the competitive advantage was generated by economies of scale, while today's new economy relies on the economics of networks which involves expanding the reach of the network, whereby the network can significantly increase its value by connecting to other networks (2007, 191). The establishment and operation of network organizations under the new management approach is driven by the desire to find modern and non-standard solutions, the development of knowledge and the support for innovation.

The concept of network organization is not clearly defined in the related literature. It can even be assumed that due to the unprecedented pace of changes, including changes in network organizations themselves, this definition will undergo further modifications. However, the description of a network structure in terms of the classical graph theory will form the basis for a whole range of concepts introduced to illustrate such structures.

3

A graph is a set of vertices that can be linked by edges in such a way that each edge ends and starts at one of the vertices (Wilson, 2000, 19). A graph (Figure 2) is written as an ordered pair G = (V, E), where:

V = {v_1, v_2, v_3,..., v_n} – is a set of n numbered vertices,

E = {e_1, e_2, e_3,...., e_m} – is a set of m edges, where each edge is a pair of vertices (V_x, V_y) connected to it, with V_x, V_y ∈ V,

where a network is understood as a set of vertices connected to each other by the edges.

Organization and management studies have adapted the graph model to their own needs, assuming that vertices are organizations that engage in collaborative relationships, and various types of flows among them form their interconnections. The general idea relevant to any type of organization is to describe it from the behavioral point of view, as a pattern of social relationships that govern relations among members of the organization (Sailer, 1978, 73-90). From a strategy perspective, networks can be considered as long-term agreements between separate organizational units that share common goals. Such actions are undertaken to maintain or enhance the competitive advantage of the resulting network organization (Jarillo, 1988).

Figure 2. A graph sketch

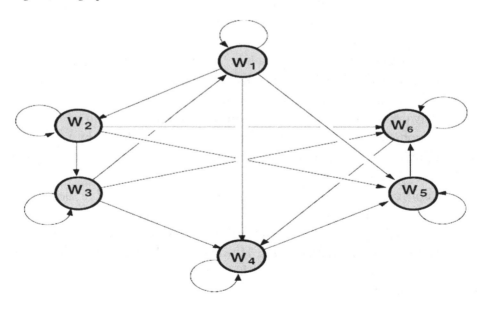

A network organization is defined by P. Colin (2004,268) as one that functions as long as possible in isolation from the traditional structure. In order to implement specific projects and defined goals, such organizations form teams. These teams are inherently impermanent since they are dissolved after project completion and the organization looks for new projects and new resources to support their implementation. The concept of network organization is often identified with the notion of virtual organization, which is also evident in P. Collin's works, where, nonetheless, the difference is that it is the structure that distinguishes a network organization, whereas a virtual organization is unique due to its operational manner that fully relies on modern ICTs (Collin, 2004, 442).

NETWORK ORGANIZATION AS THE EFFECT OF COLLABORATION OF ORGANIZATIONS: COMPETITION AND COOPERATION AS FACTORS AFFECTING THE DEVELOPMENT OF NETWORK ORGANIZATIONS

Network organizations are ones that are formed as a result of a merger of different, often independent, units. U. Kąkol and A. Kosieradzka (2014, 1-10) indicate three basic forms of inter-organizational cooperation: multi-task, multi-organizational, and multi-unit. This division makes it possible to present ways of cooperation by means of a three-dimensional matrix where the individual dimensions refer to the number of tasks being performed, the number of collaborating organizations, and the number of cooperating units within an organization (Figure 3).

This connection may be temporary, often confined to the completion of a defined, specific task. Collaborative relationships among organizations can be varied and may involve both material and non-material organizational resources (Satell, 2015; Zott & Amit, 2010). The primary goal behind the formation of such structures is to achieve synergies – faster or more efficient attainment of defined goals by pooling resources. Hence, a particular emphasis on informational connections can be seen in network structures (Bower, 2003; Maney et al, 2011).

E. Stańczyk-Hugiet indicates "business advantage", defined as economic rent, as the primary motive for forming and working within network structures (Stańczyk-Hugiet & Sus, 2012). B. Mikuła (2006) points to reduced transaction and transportation costs, often highlighted as another effect of the digital

Figure 3. Opportunities for cooperation between organizations
Source: (Kąkol & Kosieradzka, 2014, 4)

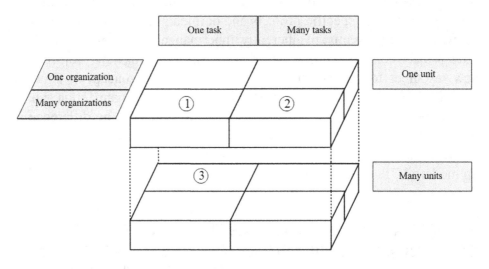

revolution, as a factor determining the existence of network organizations. He also stresses that the emergence of a new type of organization, so different from thus far existing organizational forms, requires its members to develop new management styles and models. This need, ensuing from the necessity to strengthen cooperation among self-managed expert teams and facilitate their fast and efficient communication to achieve the goals adopted by a network organization, is also emphasized in the works by D. Kirkpatrick (2011) and G. Hamel (2007). This is also a method for reducing risks and uncertainty associated with running a business, particularly under threat, since networks allow some risks to be transferred to their participants. J. Niemczyk even writes about a unique "bypass system" in the context of risk management for network organizations (2013,142). Such a system "offers a sense of greater security in the face of market competition and means more resource flexibility and lesser capital needs" (Łobejko, 2012, 10)

Organizations whose overarching goal is to build up broadly understood knowledge must have not only adequate infrastructure (including, most importantly, ICT infrastructure) but also management methods that address the needs defined by the best knowledge management standards. J. Brilman lists four types of network organizations: integrated networks, federated networks, contract networks, direct relationships networks (2002, 426-427). Collaboration and its significance is also recognized by the greatest scientific authorities. The 2009 Nobel Prize in Economics was awarded by the Bank of Sweden to Elinor Ostrom, whose research had focused on the analysis of governance of the commons. She had shown that community-based companies (cooperatives) could be more efficient than commercial firms. We also more and more often hear about the success of companies that rely on partner cooperation, i.e. cooperatives, where the overriding goal is not only to generate profits but also to derive benefits other than strictly material ones. The concept of cooperation is ever more common in everyday organizational life. It seems that it can also become a benchmark for management in the 21st century. In the business context, three main types of network organizations are mentioned (Snow et al., 1992):

1. Internal, where a large company has separate units operating as profit centers;
2. Stable, where the central company (parent company) outsources work to other associated organizations. These are long-term relationships;
3. Dynamic, meaning temporary alliances taking the form of agreements with other organizations that have (both material and non-material) resources of key importance to the agreement.

Along with the changes in the environment, models and trends in organizational studies are undergoing transformations. Cooperation processes are interesting phenomena that are developing particularly rapidly. Management theories increasingly emphasize the importance of collaboration that is becoming characteristic of today's organizational conditions. Practice shows that in many cases the traditional focus on competition is insufficient. The high dynamics of changes in today's economy forces organizations to be ready to innovate, respond to environmental requirements and keep up with successive technological challenges (Flaszewska & Zakrzewska-Bielawska, 2013, 222-223). Strong competition is ever more frequently being replaced by collaborative relationships that involve many participants, not only traditional partners but also competitors.

Cooperative action in the form of network offers organizations the opportunity to use not only their own knowledge, skills and competences but also those of their partners. For an organization, it is a chance to build or strengthen its market position. An important role in network arrangements is played by ICT systems that support the development of new models of work, communication and collaboration. In consequence, more and more various organizational networks are being established as "extended organizations." (Turbilewicz, 2013, 387). These are structures that are strongly focused on sharing, exchanging and generating knowledge, skills and competences, thereby often being the place for innovative solutions. A network is a model or metaphor that describes a system of relationships among a specified number of units. While this number can be very large in social relationships (the case of social networks), it is usually clearly defined in economic relationships. These relationships include (Easton, 1992, 3-25):

1. Links and interactions among units within the network, where links are long-term and interactions are short-term relationships;
2. The structure and position understood as the interdependence of the elements that make up the network and, as a result, the way in which they form interrelationships;
3. The process understood as a change of ties among companies as an effect of jointly implemented tasks.

The importance of informal relationships within a network organization is highlighted by J. Bower (2003), who construes a network organization as a consequence of the combination of informal networks and goals (tasks that network organization members want to complete) rather than of formal structures. G. Satell (2015) stresses this, stating that an organization can only be referred to as a network if it does not fall within the boundaries of the formal structure.

Organizations engaging in cooperation within a network choose a variety of relationships: strategic alliances, clusters, and coopetition relationships. The prime objective of each of these organizational forms is to achieve goals that are otherwise unattainable for an organization acting on its own. Nonetheless, this is not operation typical solely of business organizations. Increasingly, this form is being adopted by territorial administrations (Bartkowiak & Koszel, 2015; Matthews & Schulman, 2005) or health care (Baretta, 2008). It is also an extremely popular form of activity for scientific units (laboratories, research institutes, universities). The benefits of joint action that are most often mentioned include: access to resources that are valuable from the point of view of organizational goals, growth of innovativeness (the notion of innovation, its role in contemporary management and the place of network organizations in contributing to innovation are described in Chapter 2), reduced operating costs, and expenditure on research and development. All these factors translate into a strengthened competitive position of the organization, and may also be a method for reducing operational risk.

Coopetition is a situation where the interactions among competing organizational units include not only competitive relationships but also cooperation. It virtually concerns every aspect of the economy. It can refer to typically understood competitors, namely separate companies operating in a common sector and engaged in a classical competitive battle. We can also talk about coopetition in the case of internal organizational relationships. This is a situation in large enterprises with diversified structures, where the organizational structure is divided. In such organizations, cooperation between traditionally competing units (usually rivalry over access to resources) can be treated as intra-organizational coopetition. The underlying condition for a cooperative relationship between competitors is the opportunity to achieve shared benefits that are gained more efficiently through joint operation and the use of the resources so enriched. All such efforts are intended to take advantage of the principle of synergy, whereby the effect achieved through joint action will outweigh the sum of the effects of individual works. Among numerous definitions of coopetition that can be found in the related literature, the classical view of A. Brandenburger and B. Nalebuff, who were first to publish a book on cooperation between competitors in 1996, should be emphasized. According to those authors, coopetition is a relationship involving both cooperation and competition. Their analyses are based on the classical game theory, where the result of cooperation offers all players the chance to obtain the desired result (win-win) (Brandenburger & Nalebuff, 1996, 26). According to M. Bengtsson and S. Kock (2000, 19), coopetition is a relationship that rests on the assumption that competing organizations may interact in two ways: through competition that ensues from a conflict of interests (competitive struggle) and through simultaneous cooperation given their common interests. J. Cygler defines collaborative relationships between market rivals as "the system of flows of simultaneous and interdependent competitive and cooperative relationships between competitors retaining their organizational distinctiveness. Cooperative relationships are established in order to achieve specific strategic objectives within a specified and extended time horizon." (Cygler, 2009, 19) M. Zineldin, on the other hand, treats them as an economic situation where independent parties work together and coordinate their actions to achieve mutually agreed goals. At the same time, cooperating organizations compete both with one another and with other companies. The fundamental goal of coopetition is to establish mutually beneficial partnership relationships with other actors in the system, including competitors (Zineldin, 2004, 780).

Unlike classical market relations, meaning competition and cooperation, coopetition does not have the status of a paradigm. Depending on the market situation, it takes different forms and covers different levels of organization management. Its scope and coverage may also take other forms, hence the multitude of definitions and topologies that can be found in the related literature. There is no doubt, however, that apart from classical competition and cooperation, coopetition as an "intermediate" form combining the characteristics of both these "pure" types is often present in mutual market relations.

THE ROLE OF NETWORK ORGANIZATIONS IN THE DEVELOPMENT OF SCIENCE

Cooperation is the key word in contemporary management textbooks. It is also one of the basic notions in our everyday life. Its effects are visible in almost every aspect of it. It may occur as collaboration at virtually each and every level of the organizational and inter-organizational structure, for example, joint actions undertaken by independent countries to eliminate pollution or a common policy in a selected economic area. H. Håkansson emphasizes this fact, writing that no organization today can refer to itself as being completely isolated. The way in which contemporary organizations operate and their effects depend not only on themselves but also on the associated actors (Håkansson & Snehota. 2006). Cooperation may also involve individuals or organizational units that make targeted efforts to achieve common goals. They set up companies, non-profit organizations, organize events. Cooperation is an inseparable element in science today. Also N. Phillips undelines that „scientific collaborations, particularly international partnerships, produce some of the highest quality research. What's more, the institutions and countries that produce the most high-quality science also tend to form the strongest partnership" (Phillips, 2016).

Also universities have abandoned their historic operational model that was originally fully oriented towards education and knowledge dissemination. They are developing towards attaining educational goals through research and commercialization of their achievements (Table 1). The priority is now the ability to generate added value for the economy and society. Therefore, scientific research that represents independent "orbits" is being discontinued, and scientific activity is intended to lead to the implementation of developed

Table 1. Three generations in university development

Characteristics	University		
	Latin, medieval (1st generation)	Humboldt, research-oriented (2nd generation)	Entrepreneurial, creative (3rd generation)
Purpose	Education	Education and research	Education, research and commercialization of developed know-how
Tasks	To defend truth and dogma	To understand the laws governing the world	To generate added value for the economy and society
Method	Scholasticism	Scientific research	Scientific research and its applications
Horizon of impact	Universal, pan-Christian	National, regional (often nationalistic)	Global
Language of communication	Latin	National	English
Organizational structure	National guilds, colleges, faculties	Faculties, hierarchy	Interdisciplinary sections, networks
Management	Chancellor	Researchers	Manager

Source: (Bromski, 2013, pp. 27–28)

solutions in business (i.a. industrial) practice. Modern science is, to a large extent, a servant to the economy. Solutions that represent the purpose and the desired effect of research and research projects must be useful, implementable, and able to solve actual emerging problems.

The university has become the basic institution of contemporary economy also referred to as postindustrial economy (Cohen, 2009, 3; Etzkowitz, 2008). University research is a pillar of today's innovative economy, and university education provides companies with knowledge workers. Hence the tendency to strengthen ties between the worlds of business, public administration and science. This approach is reflected in the "triple helix" model proposed by H. Etzkowitz (2008) is shown on Figure 4.

The tendency to undertake cooperative activities and establish mutual relations between non-strictly scientific units is evident in the daily functioning of scientific institutions. The most well-known and respected higher education institutions are engaging in joint activities that result in numerous publications and projects. Indicators showing the intensity of collaboration among scientific units are included in the annual Nature Index report.

The data on international scientific cooperation that are analyzed and published annually by Science Magazine indicate 10 economies most active in cooperative activity in 2016 (Table 2)

Figure 4. Proposed new conceptual model Triple Helix Triangulation
Source: (Farinha & Ferreira, 2013, 18)

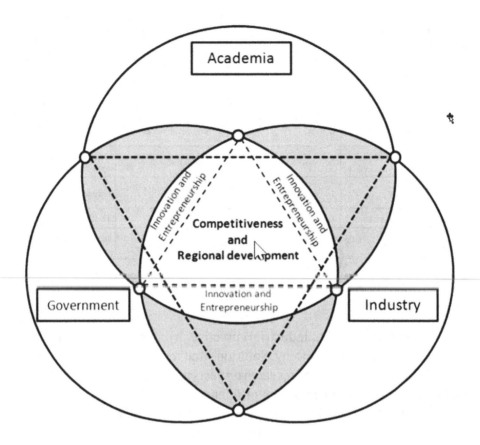

What is particularly visible there is that the cooperating institutions listed in the table include numerous representatives of the most highly esteemed universities and scientific units belonging to the world's top leaders. The main reason behind the need to take collaborative action is the fact that available resources are limited. These resources are both physical (e.g. measuring and research equipment) and human (access to specialists or so-called knowledge managers, as described, among others, by T.H. Davenport (2007, 167)). Collaborative relationships that often go beyond the boundaries of one organization or one scientific institution are an integral part of today's world.

For enterprises, it is becoming necessary to cooperate in the development of new products or in a wide range of activities called innovation development. This is determined by the high cost of research and development activities that

Table 2. State and institution co-operation based on the Nature Index 2016

No	Country	Cooperating institutions (top 5)	WFC 2014	WFC 2015	WFC change
1	USA	1. Harvard University 2. Stanford University 3. Massachusetts Institute of Technology (MIT) 4.University of California (UC Berkley) 5.University of California (UC San Diego)	18008.35	17226.51	- 4.3%
2	China	1. Chinese Academy od Sciences (CAS) 2. Peking University (PKU) 3. Nanking University (NJU) 4. Tsinghua University (TH) 5. University od Science and Technology of China (USTC)	6183.93	6481.34	+ 4.8%
3	Germany	1. Max Planck Society 2. Helmholts Association of German Research Centres 3. Ludwig Maximilian Univeristy of Munich (LMU) 4. Leibniz Association 5. Technical University of Munich (TUM)	4055.92	4086.32	+ 0.7%
4	UK	1. University of Oxford 2. University of Cambridge 3. Imperial College London (ICL) 4. University College London (UCL) 5. The University od Manchester (UoM)	3284.89	3376.22	+ 2.8%
5	Japan	1. The University of Tokyo (UTokyo) 2. Kyoto University 3. Osaka University 4. Tohoku University 5. RIKEN	3221.97	3058.12	- 5.1%
6	France	1. French National Centre for Scientific Research (CNRS) 2. Pierre and Marie Curie University (UMPC) 3. Atomic Energy and Alternative Energies Commision (CEA) 4. National Institute for Health and Medical Research (INSERM) 5. University of Paris Sud (UPSud)	2237.79	2133.05	- 4.7%
7	Canada	1. University of Toronto (UofT) 2. McGill University 3. The University of British Columbia (UBC) 4. University of Alberta (U of A) 5. University of Montreal (UdeM)	1501.97	1480.28	- 1.4%
8	Switzerland	1. Swiss Federal Institute of Technology Zurich (ETH Zurich) 2. Swiss Federal Institute of Technology in Lausanne (EPFL) 3. University of Zurich (UZH) 4. University of Geneva (UNIGE) 5. University of Basel (UB)	1300.14	1136.6	- 12.6%
9	South Korea	1. Seoul National University (SNU) 2. Korea Advanced Institute od Science and Technology (KAIST) 3. Pohang University of Science and Technology (POSTECH) 4. Yonsei University 5. Institute for Basic Science (IBS)	1182.43	1113.54	- 5.8%
10	Italy	1. National Institute for Nuclear Physics (INFN) 2. National Research Council (CNR) 3. National Institute for Astophysics (INAF) 4. Sapienza University of Rome 5. Instituto Di Ricovero e Cura a Carattere Scientifico (IRCCS)	1054.25	1065.32	+ 1.1%

Source: (Nature Index, 2016)

can represent an impassable barrier for an individual company (Merrifield, 2007). Such constraints are encountered not only by small and medium-sized enterprises, which have limited resources because of their structure. It turns out that also "giants" are more and more frequently seizing the opportunity for cooperation. Particularly spectacular cooperation is present in automotive markets where manufacturers have been working together for years to build modern cars and engines, or where, as announced by Toyota, Nissan and Honda in early 2015, they are declaring cooperation on the development of hydrogen stations for modern FCVs (Nissan, 2017). It should be noted that it is neither first nor last such cooperative relationship established in the automotive market.

Similar collaborative relationships are perceptible in virtually all sectors of today's economy. The need to develop cooperative relationships ensues from two groups of factors: internal, resulting from the specificity of the organization itself, the way of management, goals and strategies, and the organizational culture in place (Cygler, 2009). The second group of factors influencing the willingness to collaborate includes sectoral factors (Table 3) that differ considerably in the case of business and scientific entities. Regardless of the sector and the specificity of the organization itself, the development path (and the pace of market changes), however, requires organizations to be prepared for a variety of actions including cooperation.

Table 3. Impact of sectoral factors on the co-operation of enterprises and universities

Sector parametr	Impact on Business Co-Operation			Impact on University's Co-Operation		
	low	middle	high	low	middle	high
Technological advancement			+			+
Susceptibility to globalization			+	not applicable		
Intensification of competition			+		+	
Concentration of the sector's structure and intensity		+				+
Profitability and growth rate of the sector		+			+	
Entry barriers		+		+		
Confidence in supply		+		+		
The threat of substitutes	+				+	
Sector's age	+			not applicable		

Źródło: (Sobolewska, 2016, 116)

Depending on its scope, such cooperation will vary in nature. It can be pursued within a single organization when collaborative relationships are established by independent branches and departments. It may be very broad when representatives of organizations outside the sector where the initiator is active are invited to cooperate. Such a relationship occurs in the case of clusters, where the cluster consists of representatives of business, science and administration, and sometimes even representatives of the third sector, that is non-profit organizations.

Nonetheless, regardless of the scope of collaboration and the number of partners in the relationship, the optimal form of its implementation is the network structure. It is a consequence of going beyond a rigid organizational structure and related connections. Successful transfer of knowledge and technology between participants in a network organization requires mutual relationships to be built up. Such relationships will occur in the following areas: scientific partnership, research services, academic entrepreneurship, human capital mobility, commercialization of intellectual property rights, and scientific and popular science publications. Knowledge and technology are most often transferred between these entities through the appointment of a researcher/researchers acting for the university and representatives of the company. Such cooperation is governed by contracts, draws on the experience gained during the implementation of other projects, and is easier thanks to the previously developed network of contacts.

Following the achievements of adaptive learning theory, actors should collaborate in an iterative way permitting them to adapt to changes in the environment (Bjerregaard, 2009, 163). Joint projects divided into phases allow for ongoing evaluation of partial results and adjustments to joint actions in line with new information acquired. Fast feedback, both between partners and with the surrounding environment, is characteristic of networked links (Davenport, 2007, 167). Thanks to remote work with the use of information technology, communication between participants in a network is quick. Also customers/recipients of developed solutions can easily make their comments, for example, through solutions such as:

1. Intranet;
2. Knowledge bases;
3. Document management systems (dms),
4. Learning management systems (lms);

5. Audio and video tools (videoconferencing, webinars, webcasts, knowledge pills);
6. Social media tools (blogs, online forums, wiki pages, social networks, communities of practitioners).

An important aspect of a network structure is also considerable organizational flexibility of the work on projects; it is possible to increase the scope of activity, acquire new resources, propose new applications of developed solutions at any time. Flexibility is also necessary in the face of changes adversely affecting ongoing projects; then, participants in the network can concentrate on minimum-scale solutions. Collaboration within network structures fosters the development and transfer of knowledge and enhances the quality of knowledge management. Crossing the boundaries of one's own discipline and sharing experiences in one's specialization commonly results in modern, innovative solutions. Network structures also favor qualitative factors such as interpersonal relationships. They are voluntary, as opposed to those in organizations. These are sometimes very strong and long-lasting relations that significantly affect the project quality. This factor is very often the element that initiates new projects and influences the efficiency of undertaken actions. It is particularly significant when a project is carried out by highly qualified staff, so-called knowledge workers (Davenport, 2007, 119-120). The relation between the organizational structure and the communication system is described in more detail in Chapter 4 of this monograph.

In traditional organizations, a tendency to formalize and strengthen their structures can be spotted. This largely results from the need to monitor not only the work of employees themselves but also the flows of material and non-material resources among them. The flows within organizations, regardless of whether they are small or international, can be structured. In the vast majority of cases, such structuring implies squeezing those flows into organizational information and IT systems. Such cooperation may be formal, orderly and may manifest all the characteristics of a traditional organization. Such a structure is slightly harder to build in the case of a relationship that brings together many independent, structurally distinct organizations. Then, cooperation can only be partially controlled and often takes a networked form under uncertainty and instability.

THE ROLE OF NETWORK ORGANIZATIONS IN CREATING TRANSNATIONAL COOPERATION: RESEARCH RESULTS

This part of the chapter will report the results of our research conducted from April to May 2017. The research was carried out as a survey, with questionnaires addressed to a selected closed audience. Members of 14 scientific and R&D projects were invited to participate. The survey resulted in 217 correct answers the results of which will be presented below. The research was based on non-probability sampling as the respondents had participated in the implementation of network projects, that is projects involving several organizational partners, during the previous 5 years. Cooperation partners could be universities, research laboratories, companies, non-profit organizations or public administration units. Of the 14 projects, 3 were strictly scientific (meaning that only scientific organizations participated), while 6 projects involved cooperation between science and business (R&D projects resulting in the implementation and reorganization of company processes, including production). The last project group encompassed projects carried out within consortia with co-participation of non-profit organizations and public administrations, apart from business and scientific partners. Table 4 outlines the research sample and the characteristics of participants.

A large majority of survey participants were members of organizations operating within functional (36.9%) and linear (29.5%) structures; structures regarded by the related literature as more flexible and more appropriate for project tasks (organizational structures are discussed in Chapter 4) represented 18.9% for the matrix structure and 14.7% for the network structure. This may be explained by the fact that the majority of survey participants were people employed in public administration units (14%) and research centers (48.3%). Those units have a clear and well-organized structures. During the implementation of joint projects, those structures are modified towards more flexible forms (Figure 5). This concerns network structures that proved to be the dominant model in the case of project implementation (54.4% of responses indicated that projects had been carried out in a network structure). Simultaneously, the most popular form of organization adopted in participants' home units, namely the functional structure, was indicated by only 17% of respondents.

Amongst positive aspects of working in network structures, respondents mentioned the factors indicated above in this chapter: building up new, previously inaccessible resources, increasing the "bargaining power" of the

Table 4. Summary of the research sample and characteristics of the survey participants

Project Type / Project, Number	Number (Total) of Partners in the Project	International Project?	Duration of the Project (Months)	Is the Project Closed?	Number of Participants in the Survey	Number of Project Manager / Task Managers in Survey
S - 1	8	YES	36	NO	18	3
S – 2	3	NO	16	NO	16	0
S – 3	6	YES	24	YES	19	2
SB – 1	3	YES	24	NO	9	2
SB – 2	10	YES	12	NO	21	1
SB – 3	4	YES	18	YES	6	1
SB – 4	16	YES	48	NO	18	3
SB – 5	8	NO	12	YES	13	2
SB – 6	5	NO	24	NO	8	4
SBA - 1	12	YES	18	NO	24	5
SBA – 2	4	NO	12	NO	11	1
SBA – 3	8	YES	24	YES	14	2
SBA – 4	4	NO	36	YES	19	3
SBA – 5	3	YES	42	NO	21	3
Summary					217	32

project by engaging many (sometimes different) partners, and positively influencing the development of knowledge and innovation (Figure 6). Factors that are particularly strongly emphasized include the possibility of forming a cooperation network involving the establishment of new personal contacts between project partners.

Significant differences can be noted in the assessments made by the members of various participating organizations. Representatives of the world of science particularly appreciated the possibility of knowledge development (93.3% of them indicated this attribute) and access to research grants (92.4%). Businesspeople mostly pointed to the opportunity to develop personal contacts (97.5%), access to new co-workers (98.8%), and innovative products being the output of the project (92.6%). In addition to the possibility of developing a network of contacts, representatives of administration (both public and non-profit organizations) pointed to the convenience of carrying out tasks within a network (83.9%).

Figure 5. Organizational structure of the units before and during the implementation of joint design activities (N=217)

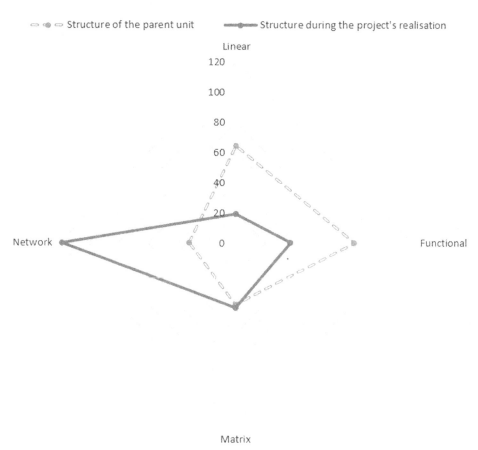

As for negative factors that are mentioned in the case of project activities undertaken within a network (Figure 7), the most frequent response was the imposition of conditions for cooperation by the party with a stronger position in the project (78%). 66% of respondents indicated the need to consult their activities with partners and diluted responsibility for project implementation (61%) as negative factors.

As in the case of the strengths of operating within network structures, some differences can be spotted in the opinions expressed by various surveyed groups about constraints. While scientific communities are reluctant to consult their activities with their partners (83% of science representatives) and administration units particularly dislike the imposition of conditions by an outside organization (76%), business representatives are most concerned

Figure 6. Positive aspects of project implementation in network structures

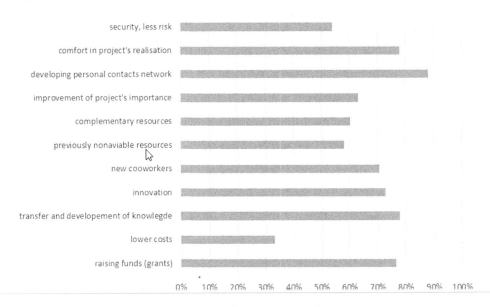

Figure 7. Negative aspects of project implementation in network structures

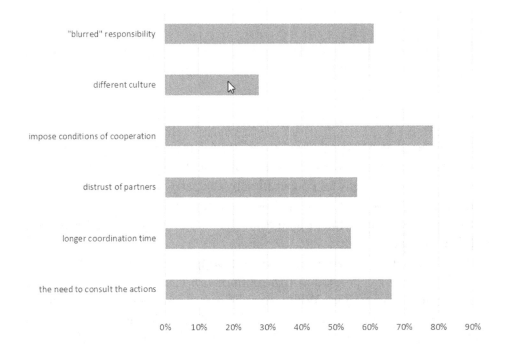

about the lack of trust in partners. This is a concern about the potential interception of know-how, knowledge and technology. This constraint is indicated by 84% of surveyed business representatives, whereas only 43% of scientists give this answer.

A lack of trust may seem particularly difficult to overcome. Nevertheless, as indicated by the interviews conducted during the survey, it is a factor that is less important to more experienced respondents. People who participate in projects implemented within a network structure for the first time regard a lack of trust in unknown partners as severe or very severe, yet trust increases as successive projects are carried out. It should be noted here that projects are undertaken with different partners and very rarely are they implemented with an already known collaborator. In the world of networks, it is necessary to open up to collaboration and to put some trust in potential co-workers. It is also very interesting to note that the greatest confidence in collaboration within networks is reported by people younger than 35 years and the oldest individuals, that is those aged 55+. As regards the former, their readiness to cooperate in networks can be explained the fact that they belong to the generation seeing the network as an intuitive tool, a companion in everyday life. In the case of the latter extreme age group of people who are open to network collaboration, it might be said that the strength of experience and an expert position built up over years prevail.

CONCLUSION

Networks have become an integral part of our daily lives. It is through networks and network links that contemporary science and industry function and develop. Nowadays, innovations are no longer produced only in high-tech, closed research and development centers. They ever more frequently form an element of cooperation among people representing very diverse social groups, fields of science and sectors of industry. They occur as a result of the interaction of these various elements, often as an effect of complex decision-making processes (this issue is analyzed in Chapter 5 of this monograph). Involvement of various individuals as well as the use of diverse tools and very different ways of looking at the problem is more and more frequently a key success factor. However, this is not an easy task, as evidenced by statistics depicting the number of failures during project implementation. Yet, this is undoubtedly another step in the development of organizations. A shift from organizations focused on improving production processes to organizations sharing processes and refining them together with the environment is what we can observe every day. If an organization opens up, not only will its organizational structures become more flexible but also, to a much greater degree, the way in which its members think will change. This is a challenge that has already been posed for today's organizations representing all possible sectors from industry to science. This is a difficult challenge that will take many years of work to meet and that requires positive experiences and changes in the way of thinking of organization members, because it is only the way of thinking that can halt the progressive and seemingly unstoppable networking process.

REFERENCES

Baretta, A. (2008). The functioning of co-opetition in the health-care sector: An explorative analysis. *Scandinavian Journal of Management, 24*(3), 209–220. doi:10.1016/j.ncaman.2008.03.005

Bartkowiak, P., & Koszel, M. (2015). Koopetycja jednostek samorządu terytorialnego, *Przegląd Organizacji, 9*, 59-76.

Bengtsson, M., & Kock, S. (2000). "Coopetition" in Business Networks – Cooperate and Compete Simulaneously. *Industrial Marketing Management, 29*(5), 411–426. doi:10.1016/S0019-8501(99)00067-X

Bielski, M. (2004). *Podstawy teorii organizacji i zarządzania*. Warszawa: Wydawnictwo C.H. Beck.

Borgatti, S., & Halgin, D. (2011). On Network Theory. *Organization Science, 22*(5), 1168–1181. doi:10.1287/orsc.1100.0641

Brandenburger, A. M., & Nalebuff, B. J. (1996). *Co-opetition. 1. A Revolutionary Mindset that Combines Competition and Cooperation. 2. The Game Theory Strategy that's Changing the Game of Business*. New York: Currency, Doubleday.

Brilman, J. (2002). *Nowoczesne koncepcje*. Warszawa: PWE.

Bromski, K. (Ed.). (2013). *Współpraca nauki i biznesu. Doświadczenia i dobre praktyki wybranych projektów w ramach Programu Operacyjnego Innowacyjna Gospodarka na lata 2007-2013*. Warszawa: Polska Agencja Rozwoju Przedsiębiorczości.

Brynjolfsson, E., & McAfee, A. (2015). *Drugi wiek maszyny*. Warszawa: MT Biznes.

Castells, M. (2007). *Społeczeństwo sieci*. Warszawa: PWN.

Cohen, D. (2009). *Three Lectures on Post-Industrial Society*. Cambridge, MA: The MIT Press.

Collin, P. (2004). *Dictionary of Business* (4th ed.). Bloomsbury Publ.

Cusin, J., & Loubaresse, E. (2018). Inter-cluster relations in a coopetition context: The case of Inno'vin. *Journal of Small Business and Entrepreneurship, 30*(1), 27–52. doi:10.1080/08276331.2017.1356158

Cygler, J. (2009). *Kooperencja przedsiębiorstw. Czynniki sektorowe i korporacyjne*. Warszawa: Oficyna Wydawnicza SGH.

Davenport, T.H. (2007). *Zarządzanie pracownikami wiedzy*. Kraków: Oficyna Wolters Kluwer business

Etzkowitz, H. (2008). *The Triple Helix. University-Industry-Government Innovation in Action*. New York: Routledge. doi:10.4324/9780203929605

Farinha, L., & Ferreira, J. J. (2013). *Triangulation of the Triple Helix: a Conceptual Framework*. Retrieved from https://www.triplehelixassociation.org

Flaszewska, S., & Zakrzewska-Bielawska, A. (2013). Organizacja z perspektywy zasobów – ewolucja w podejściu zasobowym. In Nauka o organizacji. Ujęcie dynamiczne. Warszawa: Oficyna a Woltes Kluwer.

Griffin, R. W. (1997). *Podstawy zarządzania organizacjami*. Warszawa: PWN.

Håkandon, H., & Snehota, I. (2006). No Business is an Island: The Network Concept of Business Strategy. *Scandinavian Journal of Management, 33*(3), 256–270. doi:10.1016/j.scaman.2006.10.005

Hamel, G. (2007). *The Future of Management*. Harvard Business School Press.

Jarillo, C. (1988). On Strategic Networks. *Strategic Management Journal, 6*(1), 31–41. doi:10.1002mj.4250090104

Kirkpatrick, D. (2011). *Beyond Empowerment: The Age of the Self-Managed Organization*. Morning Start, Self-Management Institute.

Kotarbiński, T. (1957). *Traktat o dobrej robocie*. Wrocław: Zakład Narodowy im Ossolińskich.

Łobejko, S. (Ed.). (2012). *Przedsiębiorstwa sieciowe i inne formy współpracy sieciowej*. Warszawa: Oficyna Wydawnicza SGH.

Matthews, J., & Schulman, A. (2005). Competitive advantage in public sector organizations: Explaining the public good / sustainable competitive advantage paradox. *Journal of Business Research*, *58*(2), 232–240. doi:10.1016/S0148-2963(02)00498-8

Merrifield, D.B. (2007). Strategic Collaborations – Essence of Survival. *Research Technology Management*, *50*(2).

Mikuła, B. (2006). *Organizacje oparte na wiedzy*. Kraków: Wydawnictwo Akademii Ekonomicznej.

Möller, K. K., & Halinen, A. (1999). Business Relationships and Networks: Managerial Challenge of Network Era. *Industrial Marketing Management*, *28*(5), 413–427. doi:10.1016/S0019-8501(99)00086-3

Nature Index. (2016). Retrieved from: https://www.natureindex.com/annual-tables/2016/institution/all/all

Niemczyk, J. (2013). *Strategia od planu do sieci*. Wrocław: Wydawnictwo Uniwersytetu Ekonomicznego we Wrocławiu.

Nissan Global. (2017). *Nissan Global website*. Retrieved from http://www.nissan-global.com/EN/NEWS/2015/_STORY/150212-03-e.html

Phillips, N. (2016, November). Nature Index 2016 Collaborations. *Nature*, *539*(7629), 17. doi:10.1038/539S1a PMID:27851720

Roig-Tierno, N., Kraus, S., & Cruz, S. (2017), The relation between coopetition and innovation/ entrepreneurship. *Review of Managerial Science, 11846*.

Sailer, L. D. (1978). Structural Equivalence: Meaning and Definition, Computation and Application. *Social Networks*, *1*(1), 73–90. doi:10.1016/0378-8733(78)90014-X

Snow, C. C., Miles, R. E., & Coleman, H. J. Jr. (1992). Managing 21st Century Network Organizations. *Organizational Dynamics*, *20*(3), 5–20. doi:10.1016/0090-2616(92)90021-E

Sobolewska, O. (2016). Kooperencja w biznesie i na uczelniach wyższych – czynniki wpływające na decyzję o relacjach współpracy. In *Dylematy rozwoju nauk o zarządzaniu. Perspektywa metodologiczna* (pp. 104–119). Toruń: Dom Organizatora.

Stańczyk-Hugiet, E., & Sus, A. (2012). Konsekwencje przynależności do sieci. In J. Niemczyk, E. Stańczyk-Hugiet, & B. Jasiński (Eds.), *Sieci międzyorganizacyjne. Współczesne wyzwanie dla teorii i praktyki zarządzania* (pp. 86–96). Warszawa: Wydawnictwo C.H. Beck.

Stoner, J. A., & Wankel, C. (1996). *Kierowanie*. Warszawa: PWE.

Wilson, R.J. (2000). *Wprowadzenie do teorii grafów*. Warszawa: Wydawnictwo Naukowe PWN.

Zineldin, M. (2004). Co-opetition: The Organization of the Future. *Marketing Intelligence & Planning*, 22(7), 780–789. doi:10.1108/02634500410568600

Chapter 2
Innovation and Its Role in Management

ABSTRACT

The chapter presents the concept of innovation and the evolution of this term in the sciences of organization and management. Innovation is a particularly important element of modern management as it contributes to creating value for the company and strengthening its market position. Regardless of the industry, every modern organization should be open to new developments; it is an expectation expressed by the recipients of products and services. The market needs to improve existing solutions and develop new products on an annual basis. Innovation is defined as the process of translating an idea or invention into a good or service that creates value for which customers will pay. By observing the actions of the most powerful organizations, it can be observed that novelties form an integral part of their business activity models. Every two years, Microsoft presents to its customers an updated version of their popular operating system, fashion houses design new clothing collections for specific dates, and Disney regularly introduces new blockbusters for the next generations of viewers. Regardless of whether an organization is global or local, the need for innovation is becoming increasingly commonplace.

DOI: 10.4018/978-1-5225-5930-6.ch002

INNOVATION: CONCEPT, CLASSIFICATION, IMPORTANCE OF INNOVATION IN CONTEMPORARY MANAGEMENT

The definition of "innovation" as found in encyclopedias can be presented in broad terms (e.g. innovation, novelties, new things introduced to the market, implementation of new technologies, founding organizations and institutions) as well as in narrow terms, describing particular branches of knowledge (e.g. product innovations, introducing new products, services and process innovations, using new and more efficient ways of obtaining these products.

Innovation is a tool for modern organizations that, if used properly, can enable businesses to take effective action in competitive markets. Oftentimes, it is innovation and the ability to create it that make it possible for companies to gain and strengthen their competitive advantage. Innovations are sometimes considered a prerequisite for running a business under current economic circumstances, or even the most valuable asset for organizations these days (Dziekoński & Chwićko, 2013). The author of one of the first definitions of the concept of innovation is J.A. Shumpeter, who already in 1912 said that it was a discontinuous performance of new combinations in the five following cases: introducing a new product or new production method; opening up a new market; obtaining a new source of raw materials or semi-finished products; and setting up a new organization for an industry (Schumpeter, 1960, 104).

Schumpeter's approach is strictly an engineering one, focusing on the productive aspect of the organization's activity. Innovation today, however, does not come down merely to production, and the original term has been conceptually expanded. We are now more often talking about innovation in terms of services, as noted by P. Kotler, who writes that innovation refers to every good that is perceived as new (Kotler, 1994, 15-28). It is also important to note there is no size requirement for innovation, as in the OECD's declaration it is the introduction of a new or significantly improved product, process, new organizational or marketing method in business practice, workplace or relations with the outside world (OECD, 2005, 46-48). E. Dworak points out that, regardless of the area of application and its scope, innovations have a number of characteristics in common. These characteristics are (Dworak. 2012):

1. **Purposefulness of Innovations:** They are not accidental and are intended to bring about beneficial changes;

2. **Scope of Changes Resulting From Innovations:** Processes, products, organization, or management methods;
3. Practical dimension and practical relevance of innovations;
4. Economic, technical or social benefits of innovations;
5. Need for a key resource in the form of economic, technical, market-related and socio-psychological knowledge;
6. Microeconomic benefits, as the result of innovation is the development of the organization;
7. Macroeconomic benefits, as the result of innovation should be positive economic effects, thus making innovation a propeller of technical progress.

The importance of innovation, along with the development of the sciences of organization and management, has also changed accordingly. This is particularly evident in the evolution of the definition of the concept in question, which dates back to Schumpeter and the early years of the 20th century and which fully corresponds to the classical trends in management, but has never ceased to advance. Today's definitions no longer mention production, but rather the implementation of innovation and the strengthening of its importance in the community (Denning & Dunham, 2010, 6). Table 1 lists some of the definitions of the concept of innovation that demonstrate a clear shift in the perception of it over the last century.

By summarizing the essence and scope of the concept of innovation in the opinions of various authors, J. Baruk indicated the common features of innovations, which are (2002, 78-79):

1. A purposeful and beneficial change in the status quo, which must find practical application;
2. Scope of changes, which involve products, services, processes, organization, management methods, market, and their result should be specific technical, economic and social benefits;
3. Fact that innovations are a means of achieving the development goals of the organization and a vehicle for technical progress (if they have a beneficial economic outcome);
4. Requirement of a specific resource of technical, market-related, economic and socio-psychological knowledge.

Table 1. Selected definitions of innovation in organization and management sciences (source: own study)

	Author	Definition
1	J.A. Schumpeter (1912)	discontinuous performance of new combinations in the five following cases: introducing a new product or new production method; opening up a new market; obtaining a new source of raw materials or semi-finished products; and setting up a new organization for an industry
2	P. Drucker (1962, 14)	Innovation is a conscious, beneficial change resulting from the need or systematic observation of the external environment
3	E. Mansfield (1968, 83)	First application of an invention
4	A.J. Harman (1971, 151-169)	Innovations consist in introducing new or significantly improved products or processes to the market
5	Ch. Freeman (1982, 7)	First commercial use of an innovative product, process, system or device (machine).
6	P. Kotler (1994, 322)	Innovation refers to any good, service or idea that is perceived by someone as new
7	W.M Grudzewski & I.K. Hejduk (2000, 138)	Any thought, behavior or thing that is new, i.e. qualitatively different from existing forms.
8	J. Fagerberg (2005, 155)	Innovations are solutions which are new and better from those currently used by people, and which influence socio-economic living conditions
9	Oslo Manual (OECD, 2015)	*"the introduction of a good or service that is a new or significantly improved with respect to its characteristics or intended uses"*
10	P.J Denning & R. Dunham (2010, 6)	*"the adoption of a new practice in a community"*

The most commonly used classification of the phenomenon of innovation is that which is specified in the Oslo Manual. This division breaks innovation down into four main types (OECD, 2008, 49-53):

1. Product innovation, or introduction of a product or service that is new or significantly improved in terms of its characteristics or applications. This includes significant improvements concerning technical specifications, components and materials, built-in software, ease of use, or other functional features.
2. Process innovation, or implementation of a new or significantly improved method of production or delivery. This category includes significant changes in technology, equipment and/or software.
3. Marketing innovation is the implementation of a new marketing method that involves significant changes in product design/structure or packaging, distribution, promotion or pricing strategies.

4. Organizational innovation is the implementation of a new organizational method in the principles adopted by the company, in the organization of the workplace or in relations with the environment.

At present, two types of businesses can be distinguished in terms of innovation. The first group is faithful to traditions and relies on proven methods and means, while the other is a company striving for success by introducing innovations and novel technologies to the market. The first group is focused on maintaining good relationships with regular customers and offering them proven products or services. This activity is to focus attention on the interior of the organization and to treat customers as passive consumers (Prahalad & Ramaswamy, 2005, 121). The actions are centered around the exchange of products between the company and the customers. Companies that use innovation in their activities are mainly geared toward the development of the organization, taking into account the dynamically changing environment they operate in. This way of running a business is reduced to developing new technologies, gaining new clients and conquering new markets. It is also important to meet customer expectations that change along with the progress of their environment. Innovation management is a highly necessary activity in modern times marked by rapid technological advancements and constant striving for change.

Factors determining innovative undertakings are (Trzepizur, 2016, 55-56):

1. Level of satisfaction of the needs of both individual and business customers.
2. Supply structure transformations, taking advantage of the market gap.
3. Possession of human and material resources, or lack thereof.
4. Competition among companies - it is much more difficult to implement a new product with a large number of competitors.
5. Market feel, i.e. knowledge of customer needs and the ability to estimate whether a product or service will be able to attract buyers.
6. Business size - large companies are more likely to implement a new product, among other factors because of their greater financial resources and properly trained staff.
7. Readiness to assume risks associated with the launch of an innovative product or service. This is due to the fact one can never be 100% sure if the novelty will prove successful and attract customers.

These factors are also evident in Figure 1, where they are elements of innovation, as represented by S. Dutta, whose view of innovation is more universal (Dutta et al, 2017, 11). While the above-mentioned factors were microeconomic, they were strictly dependent on the organization itself (the determinants of size or risk-taking were analyzed), whereas the second approach is more global. Economic factors influencing innovation are: institutions, human and financial resources, and infrastructure. Under the global economy, it is possible to map the factors in a transnational context that exceeds the boundaries of one country or region. This is one of the directions towards which modern transnational economy is headed, and this issue is the subject of analyses contained in Chapter 6. It remains clear, however, that regardless of the scope of innovation and its outreach, the ultimate product is knowledge, as highlighted in Figure 1 as well.

A company that intends introduce innovations needs to perform a number of activities in order to make the first use of new technical, technological and organizational solutions. These processes are represented by two types of linear models: *technology push model* and *market pull model*. The difference between them concerns identification of the source of innovation. In the push model, fundamental research becomes the cause of innovation, and further

Figure 1. Innovation's inputs and outputs
Source: (Dutta et al, 2017, 11)

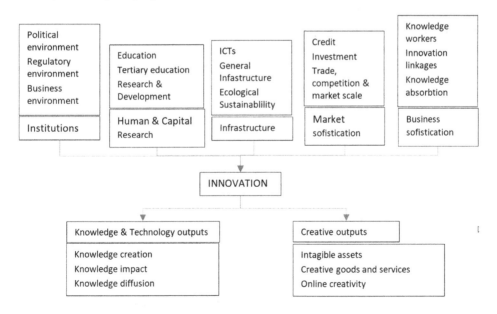

use leads to the emergence of a new product or solution. The customer is treated as a passive user who benefits from an innovation but is not expected to cooperate in creating or developing an innovative product. The second model is an innovation that has been perceived as a market need (P. Drucker, who mentions the importance of environmental and market observation in the novelty process, wrote about this type of innovation). In this case, the inspiration and immediate cause for the work on the new solution are customer expectations that remain unsatisfied. The recipient and the existing market gap are the driving force behind the commencement of basic research. This is referred to as a demand model, given that it is the unsatisfied demand that inspires an innovative action undertaken by the organization. The interaction of the supply and demand side in innovation is being increasingly discussed these days, and many examples of implemented and commercialized solutions show that the two sides of the classical market (demand and supply) are elements that, in the process of introducing innovation, should complement each other (Figure 2). The impact on innovation on the part of its potential customers is present in digital economy, a phenomenon that is more and more commonplace. This state of affairs results, to a great extent, from the ever-stronger network-based cooperation between market participants. The role of ICT infrastructure is discussed in Chapter 3 of this monograph.

Note: The adoption process, i.e. how a new product is received by its intended recipients impacts the success rate of commercialization

The literature distinguishes two key approaches to the innovation process - closed and open. The closed model is based on own resources and protection from competition. It refers to tradition and the classical trend in the sciences of organization and management. It is based on the assumption that a company is self-sufficient and that all required resources can be acquired independently, without the need for other organizational units to be involved. Limited to

Figure 2. Innovation's adoption process
Source: (Fredriksen & Knudsen, 2017)

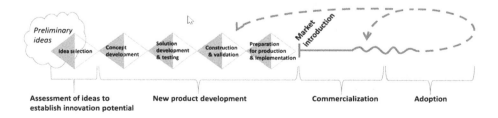

internal resources only, this process demands from organizations to be prepared to spend a considerable amount of both physical and financial resources as well as the need to update their knowledge, including through research and development. This is an important barrier that is particularly evident in the automotive industry, which P. Drucker described as *"the industry of industries"* (Drucker, 1946, 149) The sector in question leaves practically no room for small organizations and its participants are companies with strong market positions and well-known brands. These are the businesses that regularly decide to take joint action even if the partner is an active competitor. The reason for such atypical behavior is the limited access to resources, including the knowledge base, as well as the enormous costs associated with the development of further innovative solutions. In this case, there is the possibility of sharing both costs and risks between partners. Closed innovation appears to be a very convincing solution, but in a constantly changing world, fewer and fewer companies can afford to implement this strategy, precisely because of the constraints in their resources and the access to possible desirable means of production. Another factor that further hampers the choice of closed innovation is time. Even well-established organizations that do not normally struggle with access to all sorts of resources are forced by today's market to innovate as quickly as possible. Time may be another strong constraint for choosing a closed innovation strategy. The basic assumptions of closed innovation are (Kozioł-Nadolna, 2017, 297):

1. Employ properly trained staff, smart people and experienced professionals.
2. Motivate employees to come up with and develop new ideas.
3. If pursued, unique ideas offer a chance to become a pioneer in a given market.
4. Being the first to do something in a given market in most cases is identified with a win.
5. The right amount of innovation, efficient implementation and being first in the market can contribute to market leadership.
6. Controlling intellectual property to prevent employees from using the innovation to the company's disadvantage later on.

The closed innovation model relies on the use of own resources. On the other end of the spectrum, one will want to manage innovation based on an open system. In today's world, most innovations are related to interactions between individuals, organizations and the environment in which they operate. This assumes, mainly, the sharing of ideas and the acquisition of good solutions from other companies. There are also ideas that are not pursed by the organizations they originate in. In such cases, it is in many respects beneficial to sell a license for such idea to another company, which at the same time contributes to additional economic benefits for the author (Kozioł-Nadolna, 2017, 298). Open innovation is going to be described in more detail later on in this chapter.

The literature lists a number of factors influencing the readiness to undertake innovative activities by organizational units. Networking is also considered to be a particularly desirable activity, as it ensures that extra unmanaged resources (both tangible and intangible, of which human resources form important part) are guaranteed. Networking can facilitate mutual learning, which is particularly important in the innovation process. However, there is a parallel theory that does not deny the importance of network cooperation, but it speaks of the importance of close (spatially speaking) relationships and their impact on organizational creativity. It is a concept by D. Maillat and J.-C. Perrin that dates back to the early 1990s, known as *"milieu innovateur"* or *"French school of proximity"* (Maillat & Perrin, 1992). This approach highlights the essence of physical space as a condition that is essential for initiating any relationship between project members. This implies that, in line with the approach of the school of proximity, innovation requires a variety of cooperative relationships, but this cooperation should not be remote and should be implemented as closely as possible in the physical sense of the word, so that interactions between participants can take place. Organizational proximity must not, however, be reduced to a geographical dimension, as is recalled by B. Pecqueur and J.-B. Zimmermann (2004). They emphasize that proximity is made up of a number of factors, such as historical or social determinants. The French school of proximity points to three basic types of proximity, which are (Pachura, 2015, 54-55):

1. Geographic proximity, understood as the physical distance between organizations in a given territory;
2. Organizational proximity, represented by principles and formal rules of cooperation between companies;
3. Institutional proximity, taken to be a common sphere of organizational behavior, including informal behavior based in many cases on trust.

OPEN INNOVATION AS A STEP TOWARDS THE DEVELOPMENT OF NETWORK ORGANIZATIONS

The open innovation model, mentioned in the earlier section of this chapter, is based on the assumption that in today's market environment, relying solely on the resources held by the organization no longer suffices to introduce actual innovations. Coming up with innovative solutions forces organizations to use a wider pool of resources, often those that do not directly belong to the organization. The traditional (closed) approach was based on the premise that one of the key tasks of the organization was to protect its intellectual property. Organizations that are interested in developing innovative solutions, based on the assumption of the closed model, benefit from external sources, but this is primarily reduced to market observation and the acquisition of patents, licenses or copyright to specific solutions. In the open model, meanwhile, an organization – in addition out using internal sources of knowledge - is open to partnership (Figure 3). The list of partners for further cooperation can be very extensive and may include organizations (sometimes direct competitors) or scientific and research units.

Table 2 lists the reasons why organizations are increasingly shifting from the closed to open model in the innovation sphere. This requires the organization to update the way it thinks and perceives its surroundings, and sometimes it is strongly linked with the need to introduce a new organizational culture that differs from the current one. As pointed out by various authors – among them A. Osterwalder and Y. Pigneur (2012) whatever the case, it must be

Figure 3. Closed and open innovation model
Source: (Chesbrough, 2003, 36)

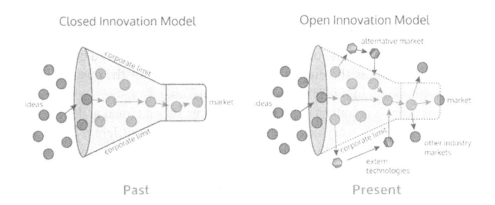

35

Table 2. Closed innovation vs Open innovation

Closed Innovation	Open Innovation
The smart people in our field work for us	Not all smart people work for us. We need to work with smart people insie and outside the company
To profit from R&D, we must discover it, develop it and ship it ourselves	Extrenal R&D can create significant value. Internal R&D is needed to claim some portion of that value
The company that gets innovation to market first will win	Building a better business model in more important than getting to market first
If we create the most and the best ideas in the industry, we will win	If we make the best use of internal and external ideas, we will win
We should control our IP, so that our competitors cannot profit from it	We should profit form other's use of our IP (license-out) and we should license in other's IP whenever it advances in our business model
We will own all results from contract research with universities	We will partner with universities to create knowledge and encourage use outside our field

Źródło: (Chesbrough, 2003)

a culture founded on a solid business model that will serve as a canvas for further activity. This model will guide developments and account for change management, which has become a determinant of modern economics. The importance of business models and their role in the development strategy for global organizations is described in detail in Chapter 6 of this monograph.

Market changes, strongly correlated with the development of information technology (ICT tools and their role in the functioning of network organizations are discussed in Chapter 3 of this monograph), have caused a gradual blurring of boundaries between the organization and its environment. Increasingly, the value provided by the company is co-created as a result of dialogue between the organization and its "stakeholders", as noted by C.K. Prahalad and V. Ramasawamy (2005, 2-5). A value co-creation model that fits perfectly into the reality of today's organization stems from interaction with the recipient of the product or service. Customers today are not just passive recipients of the finished product, but rather their co-creators (Rossi, 2011, 48). The shift that is taking place within organizations towards interaction and co-creation of values, as well as the emphasis on the importance of the customer and other stakeholders in this process, is shown in Table 3.

At this point it is important to note that cooperation aimed at developing innovative solutions, services or products, can take very different forms and cover a wide range of individuals. Such interaction can occur among active competitors, companies of similar structure operating within the same market sector – in which case one can speak of relations of cooperation. They may

Table 3. Migrating to co-creation experiences

	Traditional Exchange	**Co-Creation Experiences**
Goal of interaction	Extraction of economic value	Co-creation of value through compelling co-creation experiences, as well as extraction of economc value
Locus of interaction	Once at the end of the value chain	Repeatedly, anywhere and any time in the system
Company-customer relationship	Transaction based	Set of interactions adn transaction focused on a series of co-creation experiences
View of choice	Variety of products and services, features and functionalities, product performance, and operating procedures	Co-creation experience based on interactions across multiple chanells, options, transactions, and the price-experience relationship
Pattern of interaction between firm and customer	Passive, firm-initiated, one-in-one	Active, initiated by either firm or customer, one-on-one or one-to-many
Focus of quality	Quality of internal processes and what companies have an offer	Quality of customer-company interactions and co-creation experiences

Źródło: (Prahalad, Ramaswamy, 2004, p. 8)

also be activities that are undertaken by individuals from different sectors of the economy, differing in size, organization and management model, or organizations belonging to very diverse cultures.

The issue of the diversity of partners undertaking joint action is often discussed in the literature (Chesbrough, 2003a; Narula, 2004; Srholec, 2015). J.M. Ahn, D. Kim and S. Moon point to three basic dimensions of such cooperation: vertical, horizontal, and international (Ahn et al., 2016).

The vertical dimension occurs when innovation-oriented cooperation takes place within the organization's value chain (Figure 4), and the partners work together in the area of the common chain, e.g. the partner can be a production company and a distributor. This process may involve two separate organizations that nonetheless take joint actions motivated by the desire to improve the quality of their significant (key from the strategic standpoint) skills within the value chain. The specificity of this collaboration enables partners to make better use of their knowledge and experience, and better understand their role in customer service (Santamaria & Surocca, 2011), as

Figure 4. An example value chain for a production enterprise

well as lead to cost reduction in the value chain, which is the result of process optimization. Each of these activities enables partners not only to boost the efficiency of the process itself, but also to provide the basis for further action to improve the product and to better adapt to changing market requirements (Sadowski et al, 2003).

The horizontal dimension is when cooperation is extended to organizations outside the company's internal value chain. Competitors are also invited to cooperate with units whose key skills differ from the spectrum of the organization. Such partners may be universities, institutes or research laboratories. Inviting to cooperate completely new units allows to develop new competences. This is an extension of what is offered by vertical cooperation, where the effect of the interaction is most often the optimization of the already existing customer service process. In this case, the final product of horizontal cooperation is the extension of knowledge about the environment, acquiring new skills, or finding new areas of activity (Faems et al, 2005). One of the problems with this type of collaboration is the fact there is no natural bond that exists in the vertical dimension, i.e. there is no community of objectives, and thus ventures like this are bound to entail a significant risk of failure due to the low motivation of the participants or lack of trust between them. In the case of interactions undertaken by different organizational units, the literature points to several common determinants of such a cooperative relationship. These include (Sobolewska, 2016, 113):

1. **Duality of Interactions:** Simultaneous occurrence of competition and cooperation, made possible through the separation of areas for individual activities, and the effective coordination of activities in these separated zones of the functioning of cooperating organizations;
2. **Interdependence:** Manifested by the mutual dependence of the parties involved, and by the sharing of resources that have been contributed to the cooperation in the form of contributions, which can be of any form (material resources, know-how, skills);
3. **Duration of Relationships:** The longer the cooperation horizon, the easier cooperative relationships are established; longer cooperation also affects the quantity and variety of cooperative agreements and its internal structure;
4. **Openness:** Cooperation must be between two or more organizations, but there is no limit as to the number of parties involved; openness also applies to the markets that participate in the cooperation, as the cooperating parties need not always be direct competitors; different forms

and scopes of interaction can also be of different nature - the decisive factor in this case is the willingness of the parties and the possibility of pursuing mutual objectives.

The integration of business practice with technology and science is already an important element in the horizontal dimension. The need for such a synergy between technology and science, which more and more often goes beyond the boundaries of universities and scientific units (see Chapter 1 of this monograph), was first hinted back in the 1970s by R. Ayres, who wrote that technical needs can be satisfied only by means of complex 'systems' that are very costly and cover a number of scientific disciplines (Ayres, 1973, 273-275). The tendency to penetrate scientific disciplines so far apart is becoming increasingly noticeable. The process of integrating science and technology, as noted by Ph. Kotler (1999), takes place in three stages. The first stage is the state of distinct scientific disciplines. Each of the disciplines has its own research tools and conceptual apparatus that makes them relatively autonomous. Even if it they overlap, their meeting means the exchange of ideas and views, but it does not significantly affect the content of any of them. One step towards integration is the creation of a science that combines many disciplines. A leading example of a field that combines many different scientific disciplines is cybernetics. The third step is interdisciplinarity, where it is necessary to incorporate knowledge from various disciplines in the development of technology. The success, or failure, of such project depends directly on the results of scientific research drawn from different disciplines. It is then necessary to develop a network of links in which different organizational units are included, representing very diverse environments and disciplines. This is not by any means a simple process, as described in detail in Chapter 3 of this monograph. This project requires not only organization in the sphere of structure, but also the development of a culture of cooperation.

The international dimension stems directly from the globalization of the economy. The effects of globalization and its processes have bearing on virtually all areas of life. The following aspects of this dimension may be mentioned (Grupa Lizbońska, 1996, 48; Ohmae, 1989):

1. **Globalization of Finance:** Manifested by the deregulation of financial markets, the unprecedented mobility of capital, the rising number of mergers and acquisitions;

2. **Globalization of Markets and Competition (as Well as Business Strategies):** I.e. global economic integration, strategic alliances or globalization of factors of production;
3. **Globalization of Technology and Knowledge:** Based on the rapid flow of ideas from research centers to businesses and the development of inter-company relationships;
4. **Globalization of Lifestyles and Patterns of Consumption:** Where media plays a particularly important role in spreading the dominant lifestyle;
5. **Globalization of Legislation:** Taking place by means of the unification of the legal system, which reduces the role of national governments and parliaments (national legislation) - the resulting integration of societies.

With that being said, it must be remembered that despite the progressive process of globalization, there are still many barriers arising from local conditions and local cultures. The successful introduction of innovation requires understanding a number of determinants of the target recipient of such. "The extemt of a firm's intent to conform to different contextual disparities, such as regulations, will play a vital role in tapping into country-specific information and transforming it into valuable resources for the focal firm's innovation" (Ahn et al, 2016). It also requires a lot of investment, largely financial outlays, which demonstrates that not every economy "can afford innovation". This is

Table 4. Global Innovation Index 2017– Top 10 counties

No.	Country	Score (0-100)	Efficency Ratio
1	Switzerland	67,69	0,95
2	Sweden	63,82	0,83
3	Netherlands	63,36	0,93
4	United States of America	61,40	0,78
5	United Kingdom	60,89	0,78
6	Denmark	58,70	0,71
7	Singapore	58,69	0,62
8	Finland	58,49	0,70
9	Germany	58,39	0,84
10	Ireland	58,13	0,85

Source: (Dutta et al., 2017, xviii)

Figure 5. Movement of top 10 in GII
Source: (Dutta et al., 2017, 13)

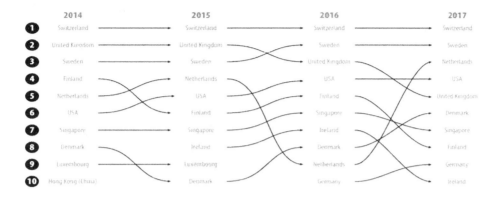

confirmed by the fact that the most innovative economies (Table 4) have been long made up by countries (Figure 5) with high income and large - according to the Global Innovation Index - expenditure on research and development.

HUMAN CENTRED INNOVATION

In open markets, the necessity to develop products that are completely dedicated to the customer and that will fully satisfy customer needs is becoming an increasingly important element determining the success or failure of a project. In this case, a success factor lies in the provision of a product fulfilling not only functional requirements but also aesthetic needs or even the need to belong to a specific user group. Actual user needs more and more frequently have to be guessed. Available data are largely historical. However, artificial intelligence algorithms can indeed be used to predict the future state. Another approach to creating innovation that is especially appreciated nowadays is the user-oriented innovation approach. Already in the first decade of the 21st century, along with the evolution of the corporate social responsibility concept (CSR is presented in chapter 5) and constantly increasing production capacities, the need to develop a completely different, modern approach to innovation was recognized (Brown, 2009). "Design thinking" is a method that seeks to combine the needs of users with the needs of their environment. This approach requires the supplier to provide not only a perfect product (or a sufficiently good one – this is considered in the chapter on decision-making) but also to involve representatives of various, often dissimilar, fields of science and technology in the design process. Such combinations make it possible to develop fully interdisciplinary solutions. This is a huge challenge for an organization that has achieved technological proficiency and established its reputation for years of its presence in the market. In this case, innovation does not stem from technical capacities of the company but from actual (or even estimated) user needs. In the case of open innovation, the need for a product was expressed by a prospective customer, whereas in "design thinking" and "human centered innovation" techniques, these needs are not obvious and should be examined in detail. To this end, the customer must be fully known since it is the human client that is a key element of the innovation design process. As much information as possible should be collected about the user of a solution. It is a stage of creative search of possibilities to satisfy customer needs, open to any and all solutions. Such openness requires that the process of seeking and developing innovations involve various people who represent diverse, often completely dissimilar perspectives. Such innovations are aimed at identifying a real problem, a difficulty faced by users. Therefore, it is necessary to open up to actual experiences and any possible solutions (Bijl-Brouwer & Dorst, 2017; Boy, 2017). Only on the basis of numerous studies, interviews, attempts to understand user needs and motives can a prototype of the solution be developed. Importantly, however, a prototype is not a solution but merely a starting point for further, deeper and more specific search. The design thinking process consists of the following stages:

1. **Empathy:** The best possible knowledge about users, their life, habits, perceptions, needs and expectations;
2. Definition of User Needs;
3. **Ideation:** Searching for most diverse, creative and even crazy solutions defined in the previous phase;
4. **Prototyping:** Developing a preliminary solution (one or several possible ways of solving the problem identified in the definition phase);
5. Prototype testing to verify its performance, expandability and usability.

NETWORK ORGANIZATIONS AS CARRIERS OF KNOWLEDGE AND INNOVATION

The definitions of innovation listed in Table 1 are only the selected samples of problems broadly contained and discussed in the literature. The previously mentioned concepts, as well as those that cannot be described in this monograph due to space constraints, emphasize the four basic elements that constitute the component of innovation. These are: the novelty of the proposed solution, its usability (possible fields of application), the success of implementation often understood as the outcome of its commercialization, and its actual use (application in practice). These elements are also shown in Figure 1 and represent the subsequent steps in the innovation process.

Using these four criteria as a determinant of innovation success of failure, an evaluation was made for the innovativeness of products coming from the projects implemented in the network structures. The structure of the sample is described in Chapter 1 of this monograph. Due to the final product of the project, the research sample was reduced to 164 questionnaires. The reason for this is the fact that the 53 questionnaires that were filled out concerned science projects and were implemented at universities and colleges, research laboratories or scientific institutes. For these projects, there was thus no need for commercialization and marketing of the final product. For the remaining 11 projects, the criterion of final product placement was fulfilled. As part of the Oslo Manual (OECD, 2008, 49-53), innovation can be divided into four groups: product, organization, process, or marketing. In the research conducted in network organizations, innovations concerned the product

Figure 6. Innovation types (OECD classification) in network organizations (N=164)

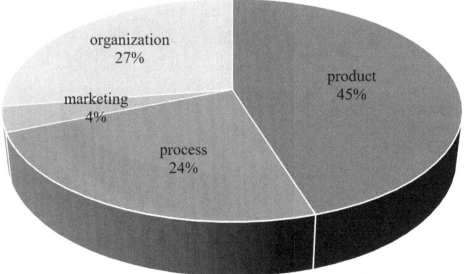

sphere. The work carried out within the network structures focused in 45% on responses concerning the product and its modernization, in 24% on the efficiency improvement and organizational reorganization, and in 27% on the organizational innovation. The remaining 4% responded to marketing innovation (Figure 6).

Figure 7 details the innovations of the project. In the case of product innovation, in accordance with the Oslo Manual, it took the form of introducing a new product (19.5% of the responses) or improving the existing product (stated by 25.6% of the respondents). In the case of process change, part of the innovation was aimed at developing a new and improved production technology (or service delivery), which concerned 14.6% of the respondents or improvement of core process activities in 21.3% of the survey responses.

It should be noted here that each of the 11 projects implemented in network structures that comprised the subject of the research in question involved different project participants. All the projects that constitute the object of analysis in this part of the monograph meet the criteria described in the earlier section of the chapter on open innovation. The projects were implemented in network structures and brought together participants from both the business

Figure 7. Type of innovation implemented in a network organization (N=164)

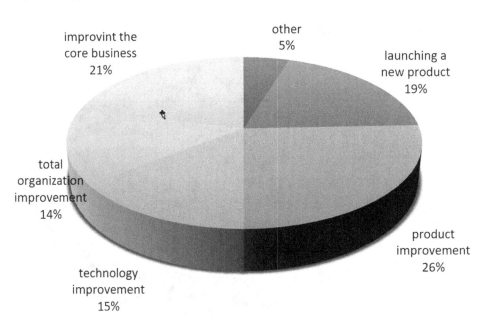

world, the public administration and the world of science. Each of the projects consisted of tasks carried out by different teams, led by separate task managers and pursuing their own pre-determined goals. Hence, within one project there were works of different type (related to product, organization or process). 6 projects (out of the 11 that made up this part of the study) can be said to have emerged as a response to the immediate needs of the market. These are innovative, demand-related projects. The other projects were supply-related projects, where the existing or current research went on to be adapted to the market requirements and commercialized in the form of a final product or as a service dedicated to the customer. In 6 projects (Figure 8), the organization that initiated the cooperation was the business side. In 3 cases, the initiator of the project was a representative of the university, and in 2 projects the initiative was on the side of the public administration.

Each of the interviewed project participants unequivocally stated that the implementation of the discussed projects, within the declared scope and time, would not be possible without networking. At the same time, a number of constraints related to the implementation of projects were declared. These limitations were not the result of technical limitations. The respondents considered the main factors affecting cooperative relations to be organizational problems and those resulting from person-to-person contacts. The majority of

Figure 8. The initiator of network collaboration

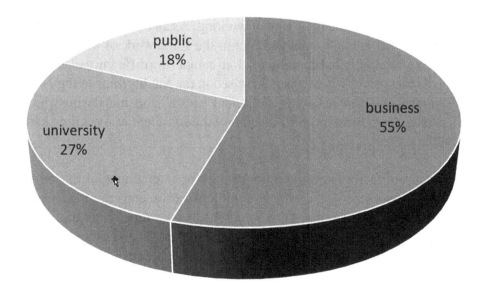

the respondents (67%) indicated the main factor to be incorrectly estimated times for each task within the project. This directly affected the synchronization of the work of each team, where the delay in one task could slow down the pace and organization of the subsequent, independent project team. The second most named factor (41%) was the problem with communication between the teams working on the project. These constraints were not the result of issues with access to technical infrastructure and instead stemmed from the following:

1. **Management in Project Teams:** A particular difficulty was noted by those teams that were centrally managed, where the person in charge of communicating with "external teams" was the task manager. In this way, project participants pointed to emerging distortions of information and, in their opinion, unnecessarily delayed verbal exchange between independent teams. However, it should also be noted that traditional, linear management of teams was in some cases justified by the importance of the project and the specificity of the cooperating unit. This was the case, among others, with projects involving sensitive data.
2. Personal limitations of project participants (unwillingness to engage in dialogue).
3. **Different Organizational Cultures:** In each of the analyzed projects, the participants came from different backgrounds and different countries. Both the local culture and the way the organization is managed (as already mentioned) significantly affects the members of the organization, their functioning and their way of working. As an example of a conflict situation that actually emerged within the framework of the analyzed projects, a very typical communication-hampering difference of attitude can be identified, where a person raised in the Mediterranean region of Europe would have his or her traditional "siesta", the time during which a parallel team from the northern part of the continent was working hard on the project.

The group of factors negatively impacting the interactions of the teams within the network projects were the administrative constraints (28% of the indications). This was due to the excessive need, in the opinion of team members, to prepare too many and too detailed reports and other types of documentation. All these, in the opinion of the project participants, unnecessary actions delayed and hindered the work on the implementation of the projects.

Figure 9. Evaluation of the project products implemented in the framework of network cooperation

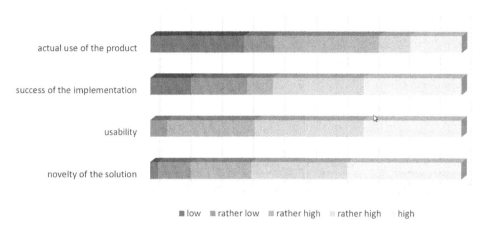

Views of innovation referred to earlier in this chapter came down to the emergence of a ncw and more practical solution that would successfully pass the commercialization process and go on to be actually used by its users. The products of the implemented projects were evaluated by their co-authors according to these criteria. The aggregate results of these evaluations, presented in the Likert five-point scale, are shown in Figure 9. It can be seen that, as far as the factor related to the practical use of project products is concerned, the respondents are fairly cautious and skeptical, they nevertheless rate the remaining three criteria very high.

KNOWLEDGE MANAGEMENT IN INNOVATIVE PROCESSES

For many organizations, the ability to create or develop innovation is inherently linked with the ability to acquire and maintain a competitive edge on the market. As already identified in the earlier section of this chapter, an organization can innovate on its own, relying solely on its own resources, or it can take obtain these resources via the process of co-creation (open innovation). Still, regardless of the innovation model, it will always rely on knowledge. At the same time, the steps involved in the creation or development of a novelty can be attributed to activities that form part of the knowledge management

Figure 10. Stages of the innovation process, knowledge management activities and desirable employee behavior
Source: (Stankiewicz & Moczulska, 2016, 38)

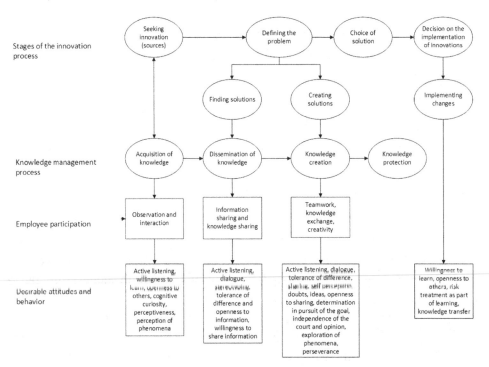

process in an organization, such as knowledge acquisition, dissemination, creation and protection (Figure 10)

The process shown in Figure 10 proves fully true for both internal and collaborative innovation. Cooperation with external entities also triggers changes in the traditional knowledge management process. Knowledge in a network organization comes from many sources, and in the course of project implementation spreads in many different directions and reaches diverse customers. The protection of knowledge will also differ in shape in the case of network organizations. The knowledge that will eventually be obtained by individual organizations and will constitute an additional product of the research project will not be the same for all collaborators. Regardless of the declared willingness to cooperate and motivation to undertake such activities, each organization ultimately seeks to protect its own *status quo*. For this reason, it should be considered natural to protect one's knowledge and develop it so as to strengthen one's competitive edge over the next period. Hence, even in the relationship of network organizations, there is a need for action of a competitive nature. Network cooperation allows for the development of knowledge, which is fully consistent with theories known in the literature. At the same time, however, it requires the collaborators to undertake actions to protect their own resources. It should be remembered that these unique resources are the basis for further projects in network structures.

A classic chain of actions that involves three basic processes - acquiring knowledge, applying it to the practical solution or project, then sharing the newly acquired knowledge that results from the implementation of project activities - becomes much more complex in the case of network organizations. The stage of acquiring knowledge is preceded by the most detailed project specification. Each of the participating organizations or project teams evaluates, prior to the implementation of the project, not only the material but also the resources of knowledge. In the case of research projects, this is the criterion for participation in project implementation. As a result of the use of knowledge in the project, the knowledge gained is developed. Newly gained knowledge will consist of explicit knowledge (documented in project papers) as well as implicit knowledge, taking the form of experience, skills, or relationships established between the participants of the project. The knowledge-sharing stage will include explicit project knowledge, whereas implicit knowledge will be left to individual members of the team or organizations involved in network cooperation.

CONCLUSION

Innovation is no longer merely an invention of a brilliant mind. Contemporary innovation has become something else. The global economy is not just about bringing together tastes and needs and creating global demand. It is also about very fast flow and access to information. Information that flows outside-in mode, i.e. from the environment toward the organization, and *vice versa*. In the latter case, one speaks of the response the organization directs to its surroundings, the recipients of its services, potential collaborators, or competitors. The stream of this information is extremely rich and, as indicated by observation, extremely difficult. Each organization will therefore strive to strengthen its market position and/or maintain it at its established level. Maintaining a competitive edge that will allow an organization sustainable operation (which is challenging in a highly turbulent market) is a must for any organization, regardless of the industry. The need to adapt their activity model to the environment expands to even the most traditional spheres of life, such as science. Contemporary universities update their curricula, develop modern tools and invite partners with whom to work on introducing innovations.

In a world that is becoming increasingly networked, it is difficult to innovate on your own. Even assuming an organization has sufficient tangible and intangible resources to develop an innovation, other factors such as time or the appropriateness of innovation must be taken into account. Improvements that made as a result of laboratory research are only part of the overall innovation market. The vast majority of modern refinements are directly correlated with market demands. The difficulty is often to understand and accurately read this need, which can be facilitated by partners. They are more and more often representatives of other fields of science or market segments. Interdisciplinarity is becoming a challenge for innovations today. Products referred to as innovative must not only fulfill the functional requirements imposed by the market but also ensure that other needs are satisfied such as comfort, ergonomics and design. Involving other units makes it possible to implement changes faster and more efficiently than competition. And it is the time factor that in many cases determines the success or failure of the project. Therefore, open innovation is being considered more and more often, as it engages distinct partners in a joint project. In the network economy, finding such collaborators and pursuing the project is no longer a significant barrier.

The element that proves particularly critical in implementing innovative network projects is knowledge. It is knowledge that constitutes an intangible asset of the organization and that is very often the reason of cooperation in the first place. For this reason, organizations are making every possible effort to protect and at the same time develop their knowledge base. In the case of joint projects, it is necessary to introduce a knowledge-sharing model, or else the complete implementation of the project may be hindered. With that being said, the organization must be able to identify the resources it would rather not share and the ones that it will want to protect.

REFERENCES

Ahn, J. M., Kim, D., & Moon, S. (2017). Determinants of innovation collaboration selection: A comparative analysis of Korea and Germany. *Innovation: Organization & Management, 19*(2), 125–145. doi:10.1080/14 479338.2016.1241152

Ayres, R. (1973). *Prognozowanie rozwoju techniki i planowanie długookresowe*. Warszawa: PWE.

Baregheh, A., Rowley, J., & Sambrook, S. (2009). Towards a Multidisciplinary Definition of Innovation. *Management Decision, 47*(8), 1323–1339. doi:10.1108/00251740910984578

Baruk, J. (2002). Innowacje, kultura innowacyjna i poziom innowacyjności przedsiębiorstw przemysłowych. *Gospodarka Narodowa, 11-12*, 84.

Baruk, J. (2013). *Innowacje jako czynnik sukcesu organizacji*. Lublin: Wydawnictwo UMCS.

van der Bijl-Brouwer, M., & Dorst, K. (2017). Advancing the strategic impact of human-centred design. *Design Studies, 53*, 1–23. doi:10.1016/j.destud.2017.06.003

Bilton, C. (2007). *Management and Creativity: From Creative Industries to Creative Management*. Oxford, UK: Blackwell Publishing.

Bissola, R., & Imperatori, B. (2011). „Organizing Individual and Collective Creativity: Flying in the Face of Creativity Clichés. *Creativity and Innovation Management, 20*(2), 77–89. doi:10.1111/j.1467-8691.2011.00597.x

Brown, T. (2009). *Change by desing*. How Design Thinking Transforms Organizations and Inspires Innovation, Harper Collins Publ.

Boy, G. A. (2017). Human-Centred Desing as an Integrating Discipline. *Journal of Systemics, 15*(1), 25–32.

Chesbrough, H. (2003). The era of open innovation. *MIT Sloan Management Review, 44*(3), 35–41.

Chesbrough, H. (2003a). *Open innovation: The new imperative for creating and profiting from technology*. Boston, MA: Harvard Business School Press.

Cropley, D. H., & Cropley, A. J. (2012). A Psychological Taxonomy of Organizational Innovation: Resolving the Paradoxes. *Creativity Research Journal, 24*(1), 29–40. doi:10.1080/10400419.2012.649234

Day, G. S. (2007). Is It Real? Can We Win? Is It Worth Doing? Managing Risk and Reward in an Innovation Portfolio. *Harvard Business Review, 85*, 110–120. PMID:18283921

Denning, P. J., & Dunham, R. (2010). *The Innovator's Way. Essential Practices for Successful Innovation*. Cambridge, MA: The MIT Press.

Drucker, P. (1946). *The Concept of the Corporation*. New York: John Day.

Drucker, P. F. (1962). *Innowacje i przedsiębiorczość. Praktyka i zasady*. Warszawa: PWE.

Dutta, S., Lanvin, B., & Wunsch-Vincent, S. (2017). *The Global Innovation Index 2017 Innovation Feeding the World*. Retrieved from https://www.globalinnovationindex.org/

Dworak, E. (2012). *Gospodarka oparta na wiedzy w Polsce: ocena, uwarunkowania, perspektywy*. Łódź: Wydawnictwo Uniwersytetu Łódzkiego.

Dziekoński, K. and Chwiećko, J. (2013). „nnowacyjność przedsiębiorstw z branży TSL. *Economics and Management*, 2, 176-193.

Faems, D., Van Looy, B., & Debackere, K. (2005). Interorganizational collaboration and innovation: Toward a portfolio approach. *Journal of Product Innovation Management*, 22(3), 238–250. doi:10.1111/j.0737-6782.2005.00120.x

Fagerberg, J. (2005). Innovation a guide to the literature. In The Oxford Handbook of Innovation. Oxford University Press.

Fleming, L. (2007). Breakthroughs and the 'Long Tail' of Innovation. *MIT Sloan Management Review*, 49, 68–74.

Frederiksen, M. H., & Knudsen, M. P. (2017). From Creative Ideas to Innovation Performance: The Role of Assessment Criteria. *Creativity and Innovation Management*, 26(1), 60–74. doi:10.1111/caim.12204

Freeman, Ch. (1982). *The Economics of Industrial Innovation*. London: F. Pinter.

George, J. M. (2007). Creativity in Organizations. *The Academy of Management Annals*, 1(1), 439–477. doi:10.1080/078559814

Grudzewski, W. M., & Hejduk, I. K. (2000). *Przedsiębiorstwo przyszłości*. Warszawa: Difin.

Grupa Lizbońska. (1996). *Granice konkurencji*. Warszawa: Poltext.

Harman, J. (1971). *The International Computer Industry. Innovation and Comparative advantage*. Cambridge, MA: Harvard University Press.

Kotler, P. (1994). *Marketing. Analiza, planowanie, wdrażanie i kontrola.* Warszawa: Wydawnictwo.

Gebethner & Ska Kozioł-Nadolna. K. (2014). Modele zarządzania innowacjami w XXI wieku. Siedlce: Zeszyty Naukowe Uniwersytetu Przyrodniczo-Humanistycznego w Siedlcach.

Maillat, D., & Perrin, J.-C. (Eds.). (1992). Entreprises innovatrices et développement territorial. Neuchâtel: GREMI, EDES

Mansfield, E. (1968). *Industrial Research and Technological Innovation.* New York: W.W. Norton.

Motyka, S. (2014). *Model kreowania systemu innowacji w przedsiębiorstwie.* Opole: Oficyna Wydawnicza Polskiego Towarzystwa Zarządzania Produkcją.

Narula, R. (2004). R&D collaboration by SMEs: New opportunities and limitations in the face of globalisation. *Technovation, 24*(2), 153–161. doi:10.1016/S0166-4972(02)00045-7

OECD. (2005). Oslo Manual: Guidelines for Collecting and Interpreting Technoogical Innovation Data (3rd ed.). Paris: OECD/Eurostat

Ohmae, K. (1989, March). The Global Logic of Strategic Alliances. *Harvard Business Review*, 143–154.

Osterwalder, A., & Pigneur, Y. (2012). *Tworzenie modeli biznesowych. Podręcznik wizjonera.* Warszawa: OnePress.

Pachura, P. (2005). Koncepcje bliskości w kształtowaniu innowacyjności. In Wyzwania współczesnego zarządzania. Tendencje w zachowaniach organizacyjnych. Toruń: Dom Organizatora.

Pecqueur, B., & Zimmermann, J.-B. (Eds.). (2004). *Economie de proximités.* Paris: Lavoisier.

Prahalad, C. K., & Ramaswamy, V. (2004). *The Future of Competition: Co-creating Unique Value with Customers.* Cambridge, MA: Harvard Business School Press Books.

Prahalad, K., & Ramaswamy, V. (2005). *Przyszłość konkurencji.* Warszawa: PWE.

Roig-Tierno, N., Kraus, S., & Cruz, S. (2017). The relation between coopetition and innovation/ entrepreneurship. *Review of Managerial Science, 11846.*

Rossi, C. (2011). Online consumer communities, collaborative learning and innovation. *Measuring Business Excellence*, *15*(3), 46–62. doi:10.1108/13683041111161157

Sadowski, B. M., Dittrich, K., & Duysters, G. M. (2003). Collaborative strategies in the event of technological discontinuities: The case of Nokia in the mobile telecommunication industry. *Small Business Economics*, *21*(2), 173–186. doi:10.1023/A:1025097812273

Santamaria, L., & Surroca, J. (2011). Matching the goals and impacts of R&D collaboration. *European Management Review*, *8*(2), 95–109. doi:10.1111/j.1740-4762.2011.01012.x

Schumpeter, J. (1960). *Teoria rozwoju gospodarczego*. Warszawa: PWN.

Sobolewska, O. (2016). Kooperencja w ujęciu współczesnych realiów współpracy strategicznej. *Studia Ekonomiczne. Zeszyty Naukowe Uniwersytetu Ekonomicznego w Katowicach*, *283*, 121-132.

Srholec, M. (2015). Understanding the diversity of cooperation on innovation across countries: Multilevel evidence from Europe. *Economics of Innovation and New Technology*, *24*(1-2), 159–182. doi:10.1080/10438599.2014.897864

Stankiewicz, J., & Moczulska, M. (2016). Zachowania pracowników sprzyjające innowacyjności organizacji w świetle badań. *Przegląd Organizacji*, *2*, 36–43.

Talke, K., & Heidenreich, S. (2014). How to Overcome Pro-Change Bias: Incorporating Passive and Active Innovation Resistance in Innovation Decision Models. *Journal of Product Innovation Management*, *31*(5), 894–907. doi:10.1111/jpim.12130

Trzepizur, P. (2016). Zarządzanie innowacjami w małych i średnich przedsiębiorstwach. *Zeszyty Naukowe Politechniki Częstochowskiej. Zarządzanie*, *24*, 55-62.

Chapter 3
Technical Infrastructure of Network Organizations

ABSTRACT

IT management infrastructure is developing hand in hand with modern information communication technologies (ICT). As with every revolution in the history of mankind, what we call the digital revolution is reflected in virtually every aspect of human life. The digital revolution has covered management to a large extent. This influence is evident both in the take-over of certain areas of organization management by IT systems (e.g., production automation) and through changes in organizations themselves and their organizational structures. Classical organizations, structured in a hierarchical fashion, follow a unique path of evolution. It began in the mid-twentieth century. What could be observed in the 1960s was the transformation of classical linear organizational relationships into matrix structures intended to streamline management and facilitate project management. The matrix structure was a breach of the classical Fayolian principle of unity of command, yet it was more flexible and conducive to cooperation compared to a traditional linear organization, thus marking the first step towards crossing borders. Nowadays, in a world where crossing organizational boundaries is becoming a management paradigm, we can more and more often see organizations of a typically network nature.

DOI: 10.4018/978-1-5225-5930-6.ch003

KNOWLEDGE MANAGEMENT IN ORGANIZATIONS

Since 1966, when P. Tomcsany first used the notion of knowledge management in the context of research works, this issue has continued to enjoy tremendous popularity both in the literature and in the everyday practice of market organizations. There are a number of knowledge management definitions. According to T.A. Stewart, knowledge management means knowing what organization members know (Edvinsson & Malone, 2001, 107). Knowledge management may also be regarded as a business process. M. Levinson defines it as a process whereby organizations create value. They use their intellectual resources to this end (Levinson, 2007). The same approach was adopted by M. Savary (1999). W.R. Bukowitz and R.L. Williams (1999, 43) take a step further, arguing that knowledge management is a process leading to wealth creation by organizations. Company management is a multi-factorial process, and processes within organizations create a value network. This network is based on interactions among organization participants resulting in both tangible and intangible values, including knowledge. Knowledge management in an organization should therefore cover all available spheres of activity, as they are either direct or indirect effect of the impact of knowledge (Figure 1).

Knowledge management, which is one of the paradigms of modern management, was signaled as early as in the 1970s by P. Drucker, who defined knowledge as the effective use of information in action and announced its increased importance in subsequent stages of economic development (Drucker,

Figure 1. Picture of the company from the perspective of knowledge
Source: (Sveiby, 2005)

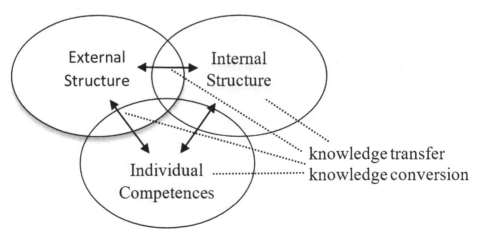

1999, 23). A similar, utilitarian approach to knowledge is represented by W. Applehans, A. Globe and G. Laugero (1999, 23), who consider knowledge as information used to solve a specific problem. Literature does not confine knowledge to the "engineering" function. Knowledge is not equivalent to information that is used effectively and efficiently, but it is also reinforced by values, experiences and skills (Tiwana, 2003, 60). According to E. Turban (1992), its shape is influenced by components such as: truth, beliefs, predictions, ideas, opinions, assumptions and expectations about the future, methodologies and know-how treated as equivalent to the other categories of company resources.

The wealth of definitions and polemics among authors is contained in one essential issue – knowledge management cannot be a one-off exercise. Knowledge management is a process that should lead towards improvement and excellence in subsequent iterations. Regardless of their industry and market position, organizations need to integrate their knowledge management processes with their management systems. Knowledge management must be a cyclical process that enables the organization to achieve its objectives effectively. This is done by transforming (both explicit and tacit) knowledge into processes such as better planning or work organization. As a result of this transformation, new organizational forms can emerge and new, non-standard tools for developing and managing knowledge resources can be found.

Knowledge management in an organization is supported today by a range of technical solutions. The opportunities offered by IT infrastructure were spotted and relatively quickly seized by companies. Already in the 1960s, IT systems were developed to support the management of production processes and warehouse management. With the technological progress, further IT solutions emerged, including those supporting the managerial staff in planning (MRP, ERP systems) and decision making (advisory systems, expert systems, business intelligence).

The development of production planning and control systems can be divided into stages at which further more advanced solutions emerged (Figure 2). These were such classes of IT solutions as (Januszewski, 2017, 153):

1. Inventory Control (IC),
2. Material Requirements Planning (MRP),
3. Closed-Loop MRP,
4. Manufacturing Resource Planning (MRPII),
5. Enterprise Resource Planning (ERP),
6. Enterprise Resource Planning (ERP II) via the Internet.

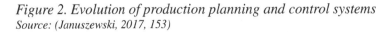

Figure 2. Evolution of production planning and control systems
Source: (Januszewski, 2017, 153)

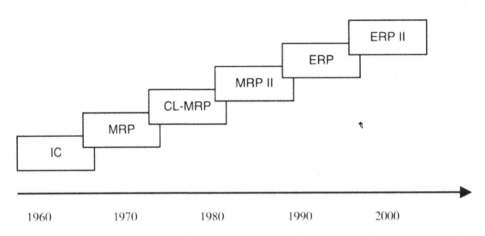

The origins of material requirement planning systems date back to the 1950s. Afterwards, the 1960s brought inventory control techniques that were based on forecasting methods and the reorder-point method. Over the years, the development of computers reduced the time needed for calculations, resulting in the MRP (Material Requirements Planning) system. The MRP methodology provides for the preparation of a production schedule in combination with necessary manufacturing materials. The system keeps track of stock levels and arranges materials with a view to the minimum possible storage time while maintaining production continuity. MRP supports production and directly related activities and areas. The next stage in the development of production control systems was MRP improvement through introducing the so-called Closed-Loop MRP, which combines MRP, MPS (Master Production Scheduling) and CRP (Capacity Resource Planning) and includes additional modules:

1. PUR (Purchasing),
2. SFC (Shop Floor Control).

In the 1980s, the functionality of CL-MRP systems was expanded with further modules: BP (Business Planning), DEM (Demand Management), SOP (Sales and Operation Planning), RRP (Resource Requirement Planning), RCCP (Rough-Cut Capacity Planning), SCR (Scheduled Receipts Subsystem). In 1989, APICS (American Production and Inventory Control Society) announced an official standard for manufacturing resource planning (MRP II) (Januszewski, 2017, 155). The next step in the evolution of production control systems was the expansion of MRP II by adding some functionalities beyond the area directly related to production. At the turn of the 1990s, financial accounting and HR and payroll modules were included, creating a new class of ERP (Enterprise Resource Planning) systems. ERP developers continue to enrich and refine their products to meet the needs of businesses. Today, enterprise management systems are often referred to as ERP II and allow online management of every aspect of an organization (web-based systems, cloud applications, mobile solutions, etc.). Moreover, the purpose of the new system class is not only to provide information services and optimize internal company processes but also to cooperate with suppliers and customers (business environment). Nowadays, the ERP II market is one of the fastest growing markets, and such systems support more and more businesses, including smaller ones.

Another group of IT solutions comprises systems for the development of communication and information networks such as intranets, web-based systems and technical solutions designed to assist the development and improvement of human resources in organizations, for example, through access to and personalization of training processes. One's position in a company determines one's information needs. Positive effects and benefits associated with knowledge management systems in enterprises include (Kłak, 2010, 162):

1. Better access to valuable knowledge;
2. Intensified innovation through free flow of ideas;
3. Increased competitive advantage;
4. Improved customer service;
5. Streamlined decision-making processes in organizations;
6. Higher revenues, given faster marketing of new products/services;
7. Shorter decision-making time by elimination of unnecessary processes, thereby reducing operating costs;
8. Optimized labor turnover by recognizing the value of employee knowledge and retaining the most valuable employees;
9. Simplified solutions to key issues;
10. Strengthened information and intellectual ties among employees;
11. Total cost reduction;
12. Emphasis on the importance of competences and continuous improvement of professional excellence in achieving company goals;
13. Shorter cycle of new product development.

Speaking of a network organization, namely one that pursues its goals beyond the traditional organizational structure, we ever more commonly mean an organization operating within the Internet environment (Colin, 2004, 268). The expansion of a global network for the exchange of information among geographically distant people and organizations was a "medium" for the development of such structures. As a result of this development and the potential offered by a distributed network, it is nowadays difficult to identify an industry or area of life that has not been "networked". Such a "networked" world was already harbingered in 2010 by the headline of The Economist's report "A world of connections." (Economist, 2010). At that time, it was also claimed that online social networks would radically change the way in which people communicate, work and play. In hindsight, it is hard to disagree with that claim. All the indications of a huge impact of computer networks on the lives of contemporary humans show that we are witnessing a real global revolution. The reach of this revolution is measured directly through the analysis of access to the Internet. In 2016, it is estimated at 46% of the world population (Internet Live Stats, 2017). This is unprecedented progress since only two decades earlier, in 1995, this access was enjoyed by less than 1% of the world population. Figure 3 illustrates the pace of development and the ever-growing number of Internet users.

SELECTED INFORMATION TECHNOLOGIES SUPPORTING THE ESTABLISHMENT AND OPERATION OF NETWORK ORGANIZATIONS

Cloud Computing

This is a term referring to the model of data access, storage and processing. Cloud integrates services, technologies and applications into a self-service tool available over the Internet. Cloud computing is a combined and resource-sharing system for gathering information from users and for virtualization. According to the widely accepted definition of the US National Institute of Standards and Technology (NIST): "Cloud computing is a model for enabling ubiquitous, convenient, on-demand network access to a shared pool of configurable computing resources (e.g., networks, servers, storage, applications, and services) that can be rapidly provisioned and released

Figure 3. Growing number of internet users
Source: (Internet Live Stats, 2017)

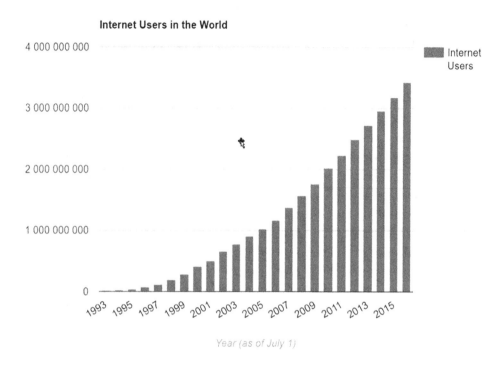

with minimal management effort or service provider interaction." (Mell & Grance, 2011, 2)

Data cloud is a processing structure based on services provided by external organizations. It is a technology that allows hiring/leasing computer infrastructure (hardware), applications and network services of professional IT companies via any computer or mobile device with Internet access. Functionality giving added value to the user is construed here as a service offered by particular software and necessary infrastructure. This means eliminating the need to install and administer the software or purchase the necessary licenses. Consumers pay a fee for using a particular service, without having to buy software or hardware. The most common cloud computing contract is not designed for a specific customer but includes a package of standardized solutions. The customer does not need to take care of the technicalities necessary for the proper operation of the service or to have knowledge of how the service is performed. There are three basic pillars of each cloud:

1. **Hardware:** Physical or virtual IT infrastructure,
2. **Software:** Operating systems, database systems or specific applications that are intended for use by the purchaser,
3. **Connectivity:** Connection to external resources, namely an access link designed for the infrastructure.

Cloud computing means that the entire burden of providing IT services (software, data, computing power) is shifted to a server where constant access is ensured for client computers. Such a procedure increases security because security does not depend on what happens to the computers of users. In addition, the speed of processes results from the computing power of the server. It is enough to log in from any electronic device with Internet access in order to start using cloud computing (Koman, 2013, 1-3).

The most common cloud services are public clouds located outside the user's enterprise. In principle, they work based on a global network of centers providing paid usage-dependent services for the general public or large industrial groups. This cloud model is applied by companies that use it to offer fast access to cheap Internet resources to different associations or individual customers. The consumer only pays attention to the most important cloud parameters and has limited possibilities to see its topography and physical infrastructure. Furthermore, consumers using this option do not have to purchase hardware, software and accompanying infrastructure that is owned and managed by the provider.

This kind of cloud computing enhances efficiency by combining demand and, consequently, by wholesale purchase of electricity and equipment. This solution reduces the unit labor cost, increasing cost savings. In addition, public cloud providers offer some services for free to increase customer interest.

Using an external cloud brings many benefits, including (Jankowski, 2015):

1. The opportunity to expand the service according to preferences;
2. Greater flexibility of solutions as the provider has access to more resources;
3. Simplicity and benefits of the installation as the costs of applications, machines and operations are covered by the service provider;
4. No over-investment in the structure as consumers pay only for what they use.

Public cloud clients have no control over resources and security of their data. Resources are actively provided online through Internet applications and services. The providers' task is to maintain the cloud on their own servers, using their own networks and storage systems allowing multiple users to use the application simultaneously. This poses a higher risk of leakage of consumer data or unintended data sharing with other customers using the allocated resources. The key here is that the provider should ensure an adequate level of security and authorization management. The cloud quality depends, to a large extent, on the service provider's ability to limit unauthorized user operations.

Examples of global public cloud providers include Goodle Apps, Microsoft Azure, Amazon Web Services, salesforce.com, and Polish providers: e24cloud. com, Beyond. These clouds are available to anyone who has an Internet connection (Mohamed, 2009).

Private cloud, also referred to as internal cloud, is a model opposite to the public cloud in terms of localization and implementation. It is intended for exclusive use by one company with multiple final users. It is managed and administered by this organization, an outside company or a combination of these two. It is also owned by them. Unlike public clouds, internal clouds are separate networks with databases using cloud computing (e.g. virtualization). Private cloud has a private (corporate) IT infrastructure that provides IT services for a defined number of users protected by a shared security system – a firewall. Private cloud management is carried out by the entity within which the cloud exists. The infrastructure is created for and allocated to one enterprise. It can be designed and managed by an external company or by the internal IT department of an organization. In the latter case, cloud computing is used by administrators at the OS (Operating System) level or above (depending on which model – SaaS, PaaS, IaaS – is implemented, as described later in this study). The company also maintains its own server rooms and people managing them.

A private cloud is appropriate for businesses keeping confidential data that should not be sent outside their systems. Such a solution does not ensure as big cost savings as a public cloud (Fuhr & Escalante, 2010, 22-34).

What exists between public and private models is hybrid clouds. They are combinations of the private and public cloud philosophies (Figure 5). The applications and infrastructure of a given entity are partially located in a private cloud, with some tasks being processed in a public cloud. This makes it possible to store confidential data in the company's system, and less protected information is transferred to public cloud servers (Antinopoulos & Gillam, 2010, 78-102).

Figure 4. The difference between private and public clouds
Source: (Herrington, 2013)

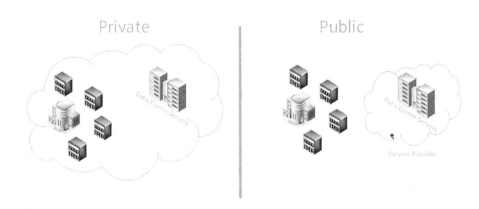

Table 1. Public and private cloud characteristics (source: elaborated by the authors)

Area	Public Cloud	Private Cloud
Infrastructure owner	External organization (cloud provider)	Company
Scalability	Unlimited and on demand	Limited by the installed infrastructure
Control and management	Only management of virtual machines, reducing the management burden	High level of resource control, better knowledge needed for resource management
Cost	Lower	High cost arising from the space, cooling, energy consumption, and the equipment itself
Efficiency	Unpredictable multitenant environment makes it difficult to guarantee the desired efficiency	Guaranteed efficiency
Security	Doubts as to data privacy	High security

Figure 5. Hybrid cloud architecture

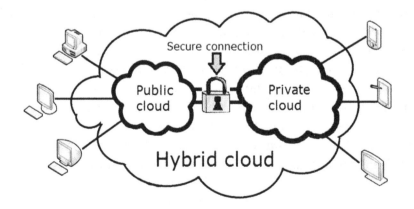

Although private clouds were considered to be the future of enterprises because of their characteristics, the hybrid model may become the most popular solution. Large corporations have already invested heavily in the infrastructure needed to manage their internal resources. Many organizations consider keeping data under their control as an important requirement for security reasons.

Hybrid cloud infrastructure includes two or more individual cloud models that form a combination of internally or externally managed servers. These clouds, interconnected by standard or proprietary technologies that allow data and application transfer, remain independent.

Such cloud models are assigned individual ID numbers, and despite being located in separate places, are joined together into one whole. Hybrid clouds offer standard or reserved access to applications and data as well as possibilities of their transfer.

Examples of global hybrid cloud providers are Amazon Web Services, IBM SmartCloud Foundation. There are also a number of smaller and local suppliers. This solution allows complete supervision to be maintained over security while using fewer critical applications in public clouds – lower costs and greater scalability as compared to private clouds (Koziar, 2013, 40-41).

According to the NIST classification, another cloud model is the community cloud. The idea behind this cloud comes from grid computing and volunteer computing paradigms.

Grid computing is based on parallel processing. The computing power of a large number of servers connected together into a cluster/grid sums up. The result is a uniform "structure" of considerable computing power.

Volunteer computing is based on grid systems. Computing resources such as CPU time or disk space are shared by computer users to perform computational tasks. This usually happens in the context of scientific projects (Jankowski, 2015).

Community cloud is a group of organizations with similar requirements that share infrastructure and increase the scale while reducing costs. It is created and managed by groups of companies that have coordinated common guidelines on security, privacy and other aspects of the cloud. Possible savings and the scale of demand are determined by the size and number of entities. A virtual data center based on virtual machines running on unused devices is another community cloud model (Balicki et al, 2012, 45-67). All types of clouds, their characteristics and differences are summarized in Table 2.

Table 2. A summary of all types of clouds (source: elaborated by the authors)

Cloud Type	
Private	Cloud infrastructure operated by the organization itself It can be managed by the organization or an outside company and can be maintained on-premise or off-premise
Public	Cloud infrastructure is available publicly or to large industrial groups and is owned by an organization selling cloud services
Hybrid	Cloud infrastructure is available publicly or to large industrial groups and is owned by an organization selling cloud services
Community	Infrastructure shared by several organizations supports a specific community of similar interests (e.g. missions, security requirements, policy) It can be managed by the organization or an outside company and can be maintained on-premise or off-premise

Big Data

We live in the age of data. Information technology and IT systems are ubiquitous, work in virtually all aspects of our lives, and force people to cooperate. We are dealing with an unprecedented amount of data. R. Smolan claims that *"the average person today processes more than one person in the 1500's did in all lifetime"* (Smolan & Erwitt, 2012, 4). Data that come from a wide variety of sources. These may be data produced by companies or government agencies but also, increasingly in the age of social networks, the information producers are individual network users whose numbers grow every minute. Each of them uses a variety of access devices, very often simultaneously. We connect to the Internet through computers, smartphones, book readers, watches, or television sets. And it is not a closed list, because it can be freely developed. Each of these devices collects and transmits huge volumes of data. Technological development has made people's data grow to the magnitudes that are not easy to measure. IDC predicts that the amount of data collected as an electronic file will increase from 4.4 zettabytes in 2013 to 44 zettabytes in 2020. Given the speed of data generation, as many data are now collected within 2 days as the total data produced by the humanity by 2003 (90% of the world data have been generated in the last few years)

Due to unprecedented data amounts, the Big Data concept has emerged. R. Smolan notes that "Big Data is an extraordinary knowledge revolution that's sweeping almost invisibly, through business, academia, government, health case, and everyday life" (Smolan & Erwitt, 2012, 3). Big Data has not been defined unambiguously; it is much more often described as what we mean by the term or, negatively, as what it is not. Table 3 summarizes selected opinions and attempts to define Big Data.

In 2001, Gartner (the then META Group) defined Big Data using three Vs: volume, variety, and velocity (Doug, 2001). Volume means that each day data are collected from various sources, including: business transactions, social media, and data sent between machines. Velocity is the need to react at an unprecedented speed, since in a situation of rapid data flow, efficient data use requires a time-coordinated response so that valuable information is not lost. Variety means that data referred to as Big Data come from various sources and in very different formats. These can be both structured (collected in databases in an orderly manner) and unstructured data. The latter are produced as a result of daily human activities and form the largest group of Big Data, estimated to be up to 90% of collected information. Most of these data are generated as e-mails, videos, posts on social media sites, telephone conversations, and search engine results.

In the last few years, Big Data has been one of the most popular topics in the business world. More and more authors see the potential of this notion (Hilbert & Lopez, 2011; Smolan & Erwitt, 2012, Phil, 2014; Couldry & Turow, 2014). With more accessible and cheaper solutions in this field, the use of Big Data is within reach of even small and medium-sized organizations. The emergence of Big Data has changed the way in which people cooperate in organizations. Managerial staff had to start working together with the IT department, since only if they work together, the value can be discovered in large datasets. Due to the variety that is so characteristic of the Big Data concept, this cooperation cannot be confined to these two employee groups,

Table 3. Selected notions describing Big Data

Notion	Author
"Big data" to analyze, the amount of which should be maximized in order to extract information values	[Cox, Ellsworth 1997]
Tendencies to search for and use business value in available ever-increasing volumes of highly volatile and complex data	SAS [www.sas.com]
Data sets that cannot be managed using current exploration methods or software tools due to the large volume and complexity of data	Fan and Bifet [2012]
A large amount of data that requires the use of new technologies and architectures, so that it is possible to extract the data value by capturing and analyzing the process – the statement by the authors	[Katal et al. 2013]
Various data generated from different sources, at a high speed and in large amounts. IBM characterizes Big Data by means of four attributes: volume, velocity, variety, and veracity.	IBM in 2013

but encompasses (or at least should encompass) all participants or stakeholders of an organization. It will involve not only organization members, traditionally understood as a group of people directly associated with the organization, namely its employees. Nowadays, people from outside the organization more and more commonly participate in developing new products and solutions. They may be those identified in project management sciences as stakeholders, that is people who can influence the status of project implementation. What is, nonetheless, important is that in the context of network organization management, this group is ever more frequently open, and the number and diversity of participants can be shaped freely according to current needs of network organizations.

Therefore, the emergence of Big Data may, to a large extent, change the approach not only to education or business but also to health, economy, medicine, and governance of countries. The effects of its implementation are ever more commonly visible in all industries. Some authors, such as U.-D. Reips and U. Matzat (2014), even write that Big Data has become fashionable. On the other hand, each, even the most attractive, solution brings about new threats. In the case of Internet solutions, there is a whole catalog of threats known collectively as cyberterrorism. As for the use of Big Data in scientific research, the scientific world sees a number of risks, apart from displaying enormous enthusiasm about the availability of previously inaccessible data resources (Boyd & Crawford, 2012; Gooding et al, 2012; Ekbia et al., 2015). These risks arise from the fear of quantity trap in scientific research, where we are not certain that the use of very large amounts of data is a guarantee of better representation of the population. This is not a new discourse as similar remarks have been made since the very beginning of research emphasizing the need to focus on data quality rather than quantity. The quantity trap "brings with it a potential negative impact upon qualitative forms of research, with digitization projects optimized for speed rather than quality, and many existing resources neglected in the race to digitize ever-increasing numbers of texts." (Gooding et al, 2012, 6-7)

Notwithstanding the reservations and doubts that are a natural effect of the emergence of a new technology and its application, Big Data is a solution that increasingly finds its place in the implementation of research projects. Such an application in medicine can be exemplified by the detection of a new strain of influenza called H1N1 in 2009. Concerns about the outbreak of a pandemic were caused by the fact that the presence of the virus had been confirmed on all permanently inhabited continents. International health organizations feared the outbreak of a pandemic. The problem was serious as there was no vaccine against the infection in 2009. Under constraints ensuing from the lack of drug, the only solution was to stop the spread of the virus. Doctors in the USA and other parts of the world were obliged to alert the Centers for Disease Control and Prevention about any newly infected person. The problem was that the information obtained from data transmitted to the monitoring organization was largely outdated. People who came to hospital often suffered from already far advanced disease, and data transmission and compilation took too long compared to the pace at which the virus was spreading. As a result, the only thing that the health care could do was to combat the already existing outbreaks of disease, and prevention possibilities were already greatly limited. The method of searching for potential outbreaks was provided by online data. Hence, engineers in the United States compared 50 million phrases that were most searched for in Google with CDC information about the spread of the regular flu virus in 2003–2008. After testing numerous models, the program developed by them detected a set of 45 phrases that, combined with a particular mathematical model, returned nearly real results. This allowed Google to predict where new outbreaks would occur at a much faster pace than the CDC could (Ginsberg et al, 2009; Wilson et al, 2009).

Internet of Things (IoT)

The term "Internet of Things" was first used by Kevin Ashton in 1999 during a presentation for Procter & Gamble on supply chain management. The Internet of Things (IoT) is a concept illustrating millions of devices that communicate with each other and share data via wired and wireless connections. Connected items such as computers, scanners, printers and phones are no longer surprising to us. A step further is to link things such as refrigerators, cars or ovens. The Internet of Things is a term describing the connection of previously unconnected electronic devices or data sources. In the network world, each of these common-place devices is a source of data; all these elements can identify, communicate, and interact with one another. The necessity of interconnectedness and the possibility of communication between these things is emphasized in the definitions of the concept in which it is mentioned "A world where physical objects are seamlessly integrated into the information network, and where the physical object can become active participants in business processes. Services are available to interact with these smart objects over the Internet, query their state and any information associated with them, taking into account security and privacy issues" (Haller et al., 2009)

The Internet of Things consists of three basic layers (shown in Figure 6): the perception layer, the network layer, and the application layer (Zhang, 2011, 4110).

1. The perception layer is the layer that collects information about objects and materials. It functions as the ends of this specific nervous system, which is responsible for receiving and gathering information about things. This layer is composed of two parts: sensors (or controllers) and short-distance transmission networks. The sensor (controller) is used to collect data and control. The short-distance transmission network sends the data collected by the sensor to the gateway or sends the commands controlling the application platform to the controller.

2. The network layer is the equivalent of the human nerve and brain and is primarily responsible for transmitting and processing information obtained by the perception layer. The network layer is formed by various private networks, the Internet, wired and wireless communication networks, network management systems, and cloud computing platforms, etc. The network layer is based on an existing communication network and the Internet. Its key technology comprises both existing communication technologies and terminal technology. Existing communication technologies include mobile communication technology, cable broadband technology, public telephone network technology, and WI-FI communications technology, etc.

3. The application layer is an interface between the Internet of Things and users (people, organizations, and other systems). It is connected with the needs of a particular industry and represents smart IoT application. It integrates ubiquitous communication networks, including the Internet, communication networks, and various access and private networks. The communication network collects, transports and processes information about objects. It is the application and business layer providing information services about mobile phones, personal computers, and other devices. It is responsible for material handling and IoT application. It ensures information collection, data mining, and decision making for all types of applications. Key technologies include smart processing of huge information, dispersed computers, middleware and information, etc.

Figure 6. IoT architecture

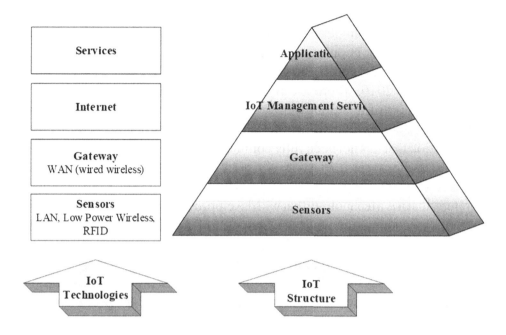

The Internet of Things primarily refers to machine-to-machine (M2M) communication and their automatic operation based on the information they share. The Internet of Things is an inherent part of the future Internet and can be defined as a self-configuring, global network infrastructure relying on standard and interoperable communication protocols. In this network, "things" have identifiers, physical attributes, virtual personalities, use interfaces, and are integrated into the web. The IoT concept assumes that "things" will be actively involved in business processes, communicating with each other and with the environment by exchanging data and information about the environment. They are also able to automatically react to events in the real world if they have already collected and examined data about it, so they can take action with or without human participation. The Internet of Things, which was originally considered a marketing curiosity, is currently one of the most interesting technologies. The Internet of Things taking the form of smart devices was mentioned by the Gartner Group as one of crucial modern technologies allowing for ever stronger synchronization of the real world with virtual reality (Gartner, 2016a). Gartner points out that what we generally mean by "smart things" includes robots, drones, and smart vehicles. This is, however, a field that will continue to develop, and smart things will be commonly used. In the annual Gartner IT Hype Cycle report (2016), this organization points to the Internet of Things as one of the most exciting new emerging technologies. IT Hype Cycle is a graphical representation of the maturity and adoption of new technologies and their impact on solving business problems. Figure 7 shows the life cycle of technology presented by Gartner in July 2016, where it can be spotted that the "Internet of Things" is at a high level of the "Innovation Trigger" segment. It is still a phase of experiments and uncertainty, but it is also the stage of the strongest development of innovation that has not been fully commercialized yet, thus providing huge room for development.

It is certain, as highlighted not only by Gartner reports but also by the related literature, that it is a technology that will develop towards collaboration, towards platforms of cooperating devices (Zhang, 2011; Chaouchi, 2010; Weber & Weber, 2010). This will be a further step towards developing a collaborative network, but this time with things as nodes. The Internet of Things aims to improve the everyday lives of users by generating information and/or processing and combining it in such a way that computers can sense, integrate, present, and react to all aspects of the physical world. Simultaneously, as suggested

Figure 7. Innovation Trigger 2016
Source: (Gartner, 2016)

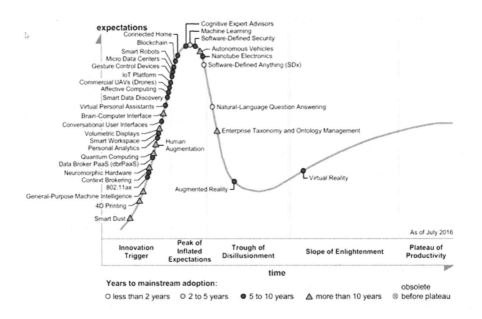

by the Gartner report, the IoT technology, which is still at an early stage of evolution, should be developed and commercialized within the next decade, meaning that we will actually witness the changes.

THE INTERNET AS CRITICAL INFRASTRUCTURE FOR THE EXISTENCE OF NETWORK ORGANIZATIONS: RESEARCH RESULTS

The ever-stronger development of ICT networks as well as increasingly widespread access to IT solutions is driving the use of these tools for knowledge management in a variety of organizations. The Internet has also followed its own evolutionary path, with Web 2.0 enjoying a particularly powerful position. It comprises all types of websites where the content is generated by users. It is a way of designing and building Internet solutions that permits the use of the so-called collective intelligence (O'Reilly, 2005). Designers' goal is to create such an infrastructure that will allow its users to independently shape the content and the form of content communication.

Thus, modern tools using ICT networks can, in the fullest possible way, support knowledge management in organizations. This is particularly important for network organizations, namely ones that very frequently rely on the Internet as the fundamental link and oftentimes as the cement that holds them together. In the case of network organizations, often geographically dispersed, it is difficult to talk about the validity of classical IT systems or integrated systems that are extremely valued in organizational monoliths (e.g. multinational corporations). Network organizations, which exist in order to achieve specific goals, will much more often apply modern IT tools. These are solutions such as:

1. Intranet;
2. Knowledge bases;
3. Document management systems (DMS);
4. Learning management systems (LMS);
5. Audio and video tools (videoconferencing, webinars, webcasts, knowledge pills);
6. Social media tools (blogs, online forums, wiki pages, social networks, communities of practitioners).

The research described in the first chapter of this monograph was supplemented by data obtained in analogous research conducted a year before (2016). The research from that year concerned Internet tools, knowledge about them and their use in the implementation of scientific projects. The participants were representatives of the scientific world. Hence, the results in Figures 1 and 2 will be based on responses to 281 questionnaires. The questions included in the survey focused on two areas. The first one pertained to the knowledge and actual use of IT tools in scientific projects within network organizations. The second area concerned the utility of particular tools in implemented projects, in particular their influence on knowledge management in those projects.

Contemporary organizations have a wide range of IT tools that can support research works. This does not only refer to highly specialized measuring or laboratory equipment that is the basic organizational resource for many projects (including projects that underpin the research described below). The subject of the research covered universal IT tools, namely those that are commonly used in different types of organizations. These solutions include known and widely used ones such as e-mail, intranet, FTP servers and Internet forums as well as solutions such as knowledge bases (more and more supported by Big Data technology) or LMS (Learning Management Systems) and knowledge pills. However, Web 2.0 solutions are more and more frequently used. These are social media (publicly available social networks, but also communities of practitioners and professional websites), video tools (videoconferencing, webcasts, webinars), and DMS (Document Management Systems) that rely on document sharing. The first group of IT tools is commonly known and used in daily work on tasks in research projects. The situation is slightly different in the case of solutions that are here jointly referred to as Web 2.0. These are tools that have a strong position among users (Figure 8). The technologies that are still the most popular in communication among project team members are e-mail and internal computer networks. As regards e-mail, no respondent said that they did not use it. On the opposite axis (very frequent use of the tool), there is the LMS technology. In the selected research group, there was no response indicating frequent use of e-learning. Its low popularity can be explained by the specificity of the research group dominated by highly qualified specialists who were knowledge creators. The opportunity to complement expertise through remote learning tools proved to be useful for those who represented centers other than academic (business, public administration).

The tools that are grouped under the collective name of social media in Figure 8 actually form a set of numerous solutions listed above. These are

Figure 8. Frequency of use of web tools during the implementation of research and research and development projects N=281

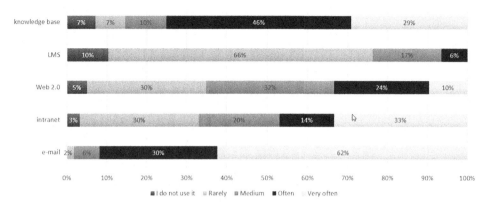

forums and blogs run by professionals, video tools that prove to be a very commonly used solution, well-known publicly available social networks, and communities of practitioners and professionals. Figure 9 presents the IT tools that were used in the implementation of the analyzed research projects. Video tools (webinars and videoconferencing) are ordinarily employed for tasks performed by geographically and organizationally dispersed project teams. Social media also provide strong support for teams as they play the role of platforms for communication among participants. Obvious popularity is enjoyed by DMS (Document Management System) solutions that allow collaborative work on documents. In this case, it is not required to be a member of a common IT network meeting the needs of network organizations that go beyond the boundaries of a single organization.

In economic sciences, utility is understood as a subjective measure of the user's preferences. It is construed as the level of satisfaction and contentment resulting from consumption of goods or services. A product or service is useful if it can meet consumer needs. The subjective nature of utility stems from a number of differences between consumers themselves and the determinants that accompany consumption processes (Rubin & Capra, 2011, 8-15). Utility is commonly measured for all types of services and products, including IT tools. The tools that were previously identified as being actually used in research work were re-examined, this time in terms of their utility in the opinion of project participants (Figure 10).

Knowledge management in organizations is an iterative process, meaning that it comprises successive stages in order to develop both knowledge

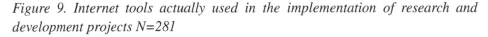

Figure 9. Internet tools actually used in the implementation of research and development projects N=281

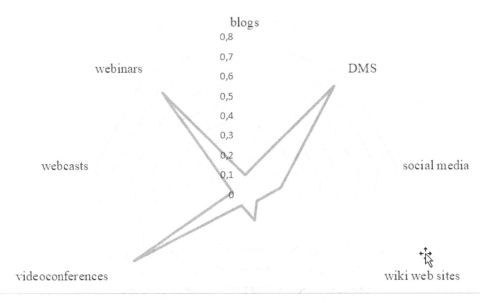

resources and the organization itself. In network organizations, it is rapid communication and exchange of resources for their efficient use that is the key factor in their establishment and growing popularity (Hamel, 2007). Today, when knowledge is the key resource and the basic building block for competitive advantages of all types of organizations, it is a priority for network entities. IT systems are responsible for supporting knowledge management in organizations. The utility of IT tools signaled earlier in the research can be referred to the subsequent stages in the classical knowledge management process (Table 4). The utility of Web 2.0 tools in knowledge localization, acquisition and development is particularly evident in the table. Knowledge bases, e-mail and classical intranet are taking over the role of knowledge archives. These are solutions that are playing a smaller and smaller role in creating new knowledge. The meaning and significance of classical IT solutions lies in support functions; the vehicles for innovation are the latest tools classified as social media.

Figure 10. Internet tools actually used in the implementation of research and development projects with their usefulness

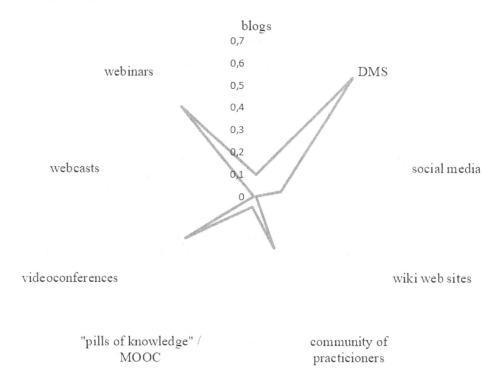

Table 4. IT tools and their importance in knowledge management processes in research and development projects

	Knowledge Location	Gaining Knowledge	Developing Knowledge	Sharing Knowledge	Using Knowledge	Behaving Knowlege
E-mail	+	+	-	+	-	+
Intranet	+	+	-	+	-	+
Knowledge bases	-	-	-	+	+	+
Videoconference	+	+	+	+	+	-
Webinar	-	+	+	+	+	-
Social media	+	+	-	+	-	-
Community of professionalns	+	+	+	+	+	-
DMS	-	+	+	+	-	-
Mind maps	+	-	+	+	-	-

REAL TIME ENTERPRISES (RTE): THE CHALLENGE FOR CONTEMPORARY ORGANIZATIONS

The world dominated by network solutions where data processing is the backbone of the economy and where continuous emphasis is placed on innovation and development requires companies to change their management fashion and adopt a completely new view of how they operate. The 21st century economy is marked by unprecedented dynamics and thus forces organizations to be open to change, to adapt their business models to the turbulent environment (Hamel, 2007). This necessity arises from a number of determinants, with the emergence of a network economy as the strongest element. It is the network economy that makes it necessary to shorten product life cycles, model and optimize existing processes in all areas of business activity, and adopt a completely new approach to customers and their needs (Wirtz et al, 2007; Turban et al, 2007). It is not easy to put all these elements in place – this requires organizations to fundamentally change not only the models and processes but also the organizational culture, as without changing this element, which is closest to the human, vigilance and readiness to change is impossible. The ideal that organizations strive for, regardless of the sphere of activity, is the formation of an organization working instantaneously, producing not only just in time, but expanding this idea to all areas of operation. The goal of today's organizations is to establish a real-time enterprise (RTE). According to A. Borg (2011), contemporary organizations should develop their ability to respond to customer and market needs and make decisions based on the information received from the environment in the shortest possible time. He also writes about the nature of sharing information within an organization, which is especially indispensable for a real-time enterprise. This is a necessary element of contemporary management, which is emphasized as important in Chapter 2 on innovation and readiness to undertake innovative actions in the practice of today's organizations. Some authors believe that the current development of ICT infrastructure and the society that habitually exploits IT technologies proves that we are indeed dealing with a real-time economy (Vasarhelyi et al, 2010; Castells, 2012; Olofson, 2013). This is an effect of the continuous data and information flow through the Internet and by means of manifold and diverse access tools, especially mobile devices that can be accessed non-stop.

The necessity to react in real time by directly responding to the expectations of the environment is becoming an element of everyday life of both organizations and people. All technological solutions, those partly listed in this chapter and other, new ones that are still being developed, arouse great hopes of analysts. According to the Accenture report of 2017, the application of artificial intelligence solutions will lead to average economic growth of 1.7% in 16 industries by 2035 (Accenture, 2017). Real-time enterprises will have to integrate their current IT systems with modern IT technologies involving AI, cloud computing and IoT solutions. As predicted by Accenture, this will generate considerable savings in intra-organizational processes and in the implementation time of these processes. Being ready to respond to information from the environment is undoubtedly an important factor in management, yet implementing such an idea requires organizations to have not only appropriate monitoring infrastructure but also procedures to respond to emergencies. Speaking of emergencies, we do not analyze public safety situations but confine ourselves to micro-crisis management. For an organization, a critical situation could involve, for example, a power failure that will prevent it from accessing a computer network or an unsuccessful promotional action that will be strongly stigmatized by the web community, which might directly affect the organization's financial performance. In such situations, monitoring the environment may allow for a rapid response and mitigation of undesirable effects of a critical situation. For real-time organizations, the ability to manage knowledge is a condition for their existence. Real-time knowledge management allows instant information to be obtained about the final product of the project and its environment.

CONCLUSION

If we analyzed the history of mankind, it would turn out that the major problem facing humans at the turn of the century was the scarcity of data, poor access to information and knowledge. Social groups or individuals that best solved this problem took over power and other resources. In the age of the Internet, the time when everyone has access to a vast amount of diverse data, the problem of access to information is theoretically losing its importance. With the wealth of data, new limitations are appearing. These are problems with the ability to indicate which data are relevant and true. In the era of the Internet and universal access to data, the problem is not access to information but the ability to find the right information. The capability of verifying and selecting data sources is growing into one of the key challenges for people and organizations today. It is becoming increasingly true that, as the amount of accumulated information increases, the possibility of using it is reduced. The hope for better use of huge data and constantly changing technologies lies in network structures. Involving new partners representing a different perspective may be an answer.

Every year, new solutions, systems and applications appear. Their quantity and variety is enormous. We also see existing solutions being ever faster replaced with newer ones that are more attractive to their users. Technology becomes outdated at a faster pace, and all kinds of predictions about the expected life of further IT solutions are doomed to failure. Despite numerous objections and doubts concerning catastrophic visions of the world dominated by thinking machines, we do not want to halt the changes in which we actively participate – as both creators and as users of subsequent generations of technological solutions. The world we live in is the world of information. And this is more and more often digital information. It is the ability to choose the right and most convenient solution from the perspective of user needs that frequently determines the success or failure of organizations. For network organizations, the prospect of getting disconnected from the global network and technological solutions would be a suicide decision. As would be the decision to give up knowledge management and development. Both knowledge and network infrastructure are a necessary condition for modern organizations to exist and grow.

REFERENCES

Accenture. (2017). *How AI boosts industry profits and innovation*. Retrieved from: https://www.accenture.com/us-en/insight-ai-industry-growth

Antonopoulos, N., & Gillam, L. (2010). *Cloud Computing. Principles, Systems and Applications.* Springer.

Applehans, W., Globe, A., & Laugero, G. (1999). *Managing Knowledge. A Practical Web-Based Approach.* Addison-Wesley.

Auksztol, J., Balwierz, P., & Chomuszko, M. (2013). *SAP. Zrozumieć system ERP*. Warszawa: PWN.

Balicki, J. (2012). *Grid and Volunteer Computing.* Gdańsk: University of Technology Press.

Borg, A. (2011). *Enterprise-grade Mobile Apps Go Global: Secure Info When and Where is Needed.* Aberdeen Group.

Boyd, D. (2010). *Privacy and Publicity in the Context of Big Data.* In WWW, Raleigh, NC.

Boyd, D., & Crawford, K. (2012). Critical Questions for Big Data. *Information Communication and Society*, *15*(5), 662–679. doi:10.1080/136 9118X.2012.678878

Brynjolfsson, E., & McAfee, A. (2015). *Drugi wiek maszyny.* Warszawa: MT Biznes.

Bukowitz, W. R., & Williams, R. L. (1999). *Knowledge Management Fieldbook.* Financial Times Prentice Hall.

Castells, M. (2012). Networks of Outrage and Hope: Social Movements in the Internet Age. *Polity*.

Chaouchi, H. (Ed.). (2010). *The internet of things: connecting objects to the web, London: ISTE.* Hoboken, NJ: John Wiley & Sons.

Colin, P. (2004). *Dictionary of Business* (4th ed.). Bloomsbury Publ.

Couldry, N., & Turow, J. (2014). Advertising, Big Data, and the Clearance of the Public Realm: Marketers' New Approaches to the Content Subsidy. *International Journal of Communication*, *8*, 1710–1726.

Cox, M., & Ellsworth, D. (1997). *Managing Big Data for Scientific Visualization, ACM SIGGRAPH '97 Course #4, Exploring Gigabyte Datasets in Real-Time: Algorithms.* Los Angeles, CA: Data Management, and Time-Critical Design.

Doug, L. (2001). Data Management: Controlling Data Volume, Velocity, and Variety. In *Application Delivery Strategies*. META Group (currently with Gartner).

Drucker, P. (1999). *Społeczeństwo pokapitalistyczne.* Warszawa: PWN.

Economist. (2010). A world of connections. *The Economist.* Retrieved from: http://www.economist.com/node/15351002

Edvinsson, L., & Malone, M. S. (2001). *Kapitał intelektualny.* Warszawa: Wydawnictwo Naukowe PWN.

Ekbia, H., Mattioli, M., Kouper, I., Arave, G., Ghazinejad, A., Bowman, T., ... Sugimoto, C. R. (2015). Big data, bigger dilemmas: A critical review. *Journal of the Association for Information Science and Technology*, *66*(8), 1523–1545. doi:10.1002/asi.23294

Fan, W., & Bifet, A. (2012). Mining big data: current status, and forecast to the future. ACM SIGKDD Explorations Newsletter, 14(2), 1-5.

Furht, B., & Escalante, A. (2010). *Handbook of Cloud Computing*. Springer. doi:10.1007/978-1-4419-6524-0

Gartner. (2016a). *Hype Cycle Special Report*. Retrieved from: http://www.gartner.com/newsroom/id/3412017

Gartner. (2016b). *Top 10 Strategic Technology Trends for 2017*. Retrieved from: http://www.gartner.com/smarterwithgartner/gartners-top-10-technology-trends-2017/

Ginsberg, J., Mohebbi, M. H., Patel, R. S., Brammer, L., Smolinski, M. S., & Brilliant, L. (2009). Detecting influenza epidemics using search engine query data. *Nature*, *457*(7232), 1012–1014. doi:10.1038/nature07634 PMID:19020500

Haller, S., Karnouskos, S., & Schroth, C. (2009). The internet of things in an enterprise context. *Proceedings: The First Future Internet Symposium*, 14-28. 10.1007/978-3-642-00985-3_2

Hamel, G. (2007). *The Future of Management*. Harvard Business School Press.

Herrington, B. (2013). *How to Compare the Private and Public Cloud*. Retrieved from: http://blog.inin.com/how-to-compare-the-private-and-public-cloud

Hilbert, M., & López, P. (2011). The World's Technological Capacity to Store, Communicate, and Compute Information. *Science*, *332*(6025), 60–65. doi:10.1126cience.1200970 PMID:21310967

Internet Live Stats. (2017). Retrieved from: http://www.internetlivestats.com/internet-users/

Jankowski, P. (2015). *Czy warto przechowywać dane w chmurze?* Retrieved from: http://www.benchmark.pl/testy_i_recenzje/jaki-dysk-w-chmurze-wybrac-korzysci-z-chmury.html

Jankowski, P. (2016). *Wszystko o usłudze Dropbox*. Retrieved from http://www. komputerswiat.pl/centrum-wiedzy-konsumenta/uslugi-online/wszystko-o-chmurach/wszystko-o-usludze-dropbox.aspx

Januszewski, A. (2017). *Funkcjonalność informatycznych systemów zarządzania*. Warszawa: Wydawnictwo Naukowe PWN.

Katal, A., Wazid, M., & Goudar, R. H. (2013). Big Data: Issues, Challenges, Tools and Good Practices. *Sixth International Conference on Contemporary Computing (IC3)*, 404-409. 10.1109/IC3.2013.6612229

Kisielnicki, J. (2014). *Zarządzanie i informatyka*. Warszawa: Placet.

Kłak, M. (2010). *Zarządzanie wiedzą we współczesnym przedsiębiorstwie*. Kielce: Wydawnictwo WSEiP.

Koman, J. (2013). ABC ekonomiki chmury obliczeniowej w dużej, średniej i małej firmie. *Biznes benchmark magazyn*, *9*, 1-3.

Korolov, M. (2013). *15 most powerful Big Data companies*. Network World.

Koziar, N. (2013). Infrastruktura chmury. *Biznes benchmark magazyn*, (3), 40-41.

Levinson, M. (2007). Knowledge Management Definition and Solutions. *CIO Magazine*. Retrieved from: http://www.cio.com/article/2439279/enterprise-software/knowledge-management-definition-and-solutions.html

Maney, K., Hamm, S., & O'Brien, J. M. (2011). *Making the World Work Better*. IBM Press.

Mayer-Schonberger, V., & Cukier, K. (2014). Big Data. *Rewolucja, która z.*

Mell, P., & Grance, T. (2011). *The NIST Definition of Cloud Computing*. Recommendations of the National Institute of Standards and Technology. Retrieved from http://nvlpubs.nist.gov/nistpubs/Legacy/SP/nistspecialpublication800-145.pdf

Mohamed, A. (2009). A History of Cloud Computing. *ComputerWeekly*. Retrieved from: http://www.computerweekly.com/feature/A-history-of-cloud-computing

O'Reilly, T. (2005). *What is Web 2.0. Designing Patterns and Business Models for the Next Generation of Software*. Retrieved from: http://www.oreilly.com/pub/a/web2/archive/what-is-web-20.html

Olofson, C. W. (2017). *Worldwide Relational Database Management Systems 2013-2017*. Retrieved from: https://www.idc.com/getdoc.jsp?containerId=US40428116

Paharia, R. (2014). *Lojalność 3.0. Jak zrewolucjonizować zaangażowanie klientów i pracowników dzięki big data i grywalizacji*. Warszawa: MT Biznes.

Phil, S. (2014). *The visual organization: data visualization, big data, and the quest for better decisions*. John Wiley & Sons, Incorporated.

Probst, G., Raub, S., & Romhardt, K. (2002). *Zarządzanie wiedzą w organizacji*. Kraków: Oficyna Ekonomiczna.

Rajaraman, A., & Ullman, J. (2011). *Map-Reduce and the New Software Stack. Mining of Massive Datasets*. Cambridge University Press. Retrieved from: http://infolab.stanford.edu/~ullman/mmds/ch2.pdf

Reips, U.-D., & Matzat, U. (2014). Mining "Big Data" using Big Data Services. *International Journal of Internet Science, 1*(1), 1–8.

Rubin, P. H., & Capra, M. (2011). The evolutionary psychology of economics. In Applied Evolutionary Psychology. Oxford University Press. doi:10.1093/acprof:oso/9780199586073.003.0002

Savary, M. (1999). Knowledge management and competition in the consulting industry. *California Management Review, 41*(2), 95–107. doi:10.2307/41165988

Smolan, R., & Erwitt, J. (2012). *The human face of Big Data*. EMC Corporation.

Stoner, J. A., & Wankel, Ch. (1996). *Kierowanie*. Warszawa: PWE.

Sveiby, K. E. (2005). Dziesięć sposobów oddziaływania wiedzy na tworzenie wartości. *E-mentor, 2*(9), 49–52.

Tiwana, A. (2003). *Przewodnik po zarządzaniu wiedzą. E-biznes i zastosowania CRM*. Warszawa: Placet.

Turban, E. (1992). *Expert Systems and Applied Artificial Intelligence*. Prentice Hall College, Macmillan.

Turban, E., & Sharda, R. (2007). *Business Intelligence*. Prentice Hall.

Vasarhelyi, M. A., Teeter, R. A., & Krahel, J. P. (2010). Audit Education and the Real-Time Economy. *Issues in Accounting, 25*(3), 1–20.

Weber, R. H., & Weber, R. (2010). *Internet of Things. Legal Perspectives.* Berlin: Springer. doi:10.1007/978-3-642-11710-7

Wilson, N., & Mason, K. (2009). Interpreting "Google Flu Trends" data for pandemic H1N1 influenza: The New Zealand experience. *Euro Surveill, 14*(44). Retrieved from: http://www.eurosurveillance.org/ViewArticle. aspx?ArticleId=19386

Wirtz, B. W., Mathieu, A., & Schilke, O. (2007). Strategy in High-Velocity Environments. *Long Range Planning, 40*(3), 295–313. doi:10.1016/j. lrp.2007.06.002

Wong, P. C., & Thomas, J. (2004). *Visual analytics. In IEEE Computer Graphics and Applications* (pp. 20–21). IEEE.

Zhang, Y. (2011). Technology framework of the Internet of Things and its application. *International Conference on Electrical and Control Engineering,* 4109-4112. 10.1109/ICECENG.2011.6057290

Zikopoulos, P., & deRoos, D. (2013). *Harness the Power of Big Data: The IBM Big Data Platform.* McGraw-Hill.

Chapter 4
Communication in Traditional and Network Organizations

ABSTRACT

In this chapter, the authors compare communication systems of network and traditional (hierarchical) organizations. They will also examine the factors that affect smooth functioning of modern communication systems. They are particularly interested in the role of ICT as a determinant of the operation of such systems. Communication systems of organizations depend on many factors. The crucial ones are the purposes that these systems are to serve and the resources allocated to them. Communication systems are analyzed in numerous works of literature on organization and management sciences. Those interested in these issues can be recommended to read the multi-volume Encyclopedia of Information Technology and Communication edited by M. Khosrow-Pour (2017). In order to present communication in organization management, the authors will employ a model approach that outlines managerial communication in a simplified way. The selection of analyzed sources is anchored in the study of literature and research reports as well as in their own experience as designers and researchers. The authors also point out that the issues highlighted in this chapter are elaborated on in other parts of this monograph.

DOI: 10.4018/978-1-5225-5930-6.ch004

MODELS OF COMMUNICATION SYSTEMS AND THEIR ELEMENTS

The cybernetic model of contemporary communication was first presented in the literature by Shannon and Weaver. The Shannon–Weaver model of communication has been called the *"mother of all models"* (Woods & Hollnagel, 2005). It is considered when many senders and receivers exist and when information channels are broadened and extended. Thereby, we can prove that when analyzing the extended model presented in the figure, the risk of losses and changes in information being sent increases. This model justifies the assertion that IT systems allow the information transmission process to be organized in such a way that we reduce the possibility of noise as an element of information loss.

The notion of communication model is applied in various contexts, and research into communication is multidisciplinary. Such research is associated with the theory and research in the fields of psychology, economics, sociology, political science, and sciences. As claimed by W.F. Eadie and R. Goret (2013), despite its deep roots in research, communication has, however, a relatively short history as a scientific discipline. The most extensive ancient texts that address issues related to communication were written by Greek and Roman societies. The importance of communication as well as the ability to communicate effectively was an indispensable part of the education of the ancient man. Rhetoric, or the art of discourse, was also a component of classical education in later centuries, representing the "trivium" of liberal arts (Lat. *septem artes liberales*), together with dialectics and grammar.

Thus, until the 19th century, the art of communication was an essential element to be learned by every educated man.

In our discussion, we focus on management and organization. We analyze the processes pertaining to two key management issues, namely efficient

Figure 1. Communication model according to Shannon & Weaver
Source: (Shannon & Weaver, 1948)

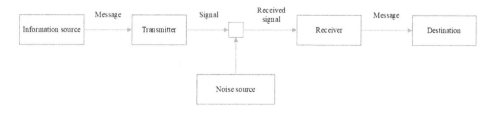

transmission of reliable and up-to-date information and knowledge to all authorized members of the organization. Today's communication system fulfills the following two tasks:

1. Transfer of information;
2. Transfer of knowledge, including the transformation of information into knowledge by means of BI information systems.

The primary objective of communication projects is to design such a process of information and knowledge transfer within an organization that will satisfy user needs. At the same time the implementation process should be effective, smooth and efficient. In the related literature, similar aspects of communication processes were addressed by many researchers. One of the broadest analyses was performed by already mentioned William F. Eadie and Robin Goret (2013). They examined the communication model in the following dimensions: the element influencing public opinion and culture, the language used, the way of transmitting information, and the developer of relationships. A comprehensive monograph on the theory of communication processes was offered by Werner J. Severin and James W. Tankard (2014). They point out that the assessment of communication processes within an organization is multidisciplinary (interdisciplinary) and requires an analysis of many of its aspects. This fact was noted by B. Bloom (1956). His works outlining so-called Bloom's taxonomy contain a classification of learning objectives. Although developed in 1956, the taxonomy is still a very useful tool. In addition to the classification of learning objectives, Bloom's taxonomy shows a model of the development of knowledge transfer systems. It consists of three domains: cognitive, affective, and psychomotor. The cognitive domain is particularly important, comprising both knowledge and intellectual skills.

In 2007, Andrew Churches developed Bloom's digital taxonomy, aiming to combine the cognitive sphere with 21st-century digital skills. The modified approach is not based on learning processes, as was the case under the classical approach, but relies on activities that are undertaken within contemporary cognitive processes. The new perspective on the classical approach to communication focuses on activities such as remembering, understanding, applying knowledge, analyzing and evaluating, and creating new solutions based on available information and tools (Figure 3).

The analysis of the communication system concerns both intra-organizational processes and those between the organization and its environment. The approach herein is an extension of the Shannon-Weaver model. The starting

Figure 2. Bloom's taxonomy

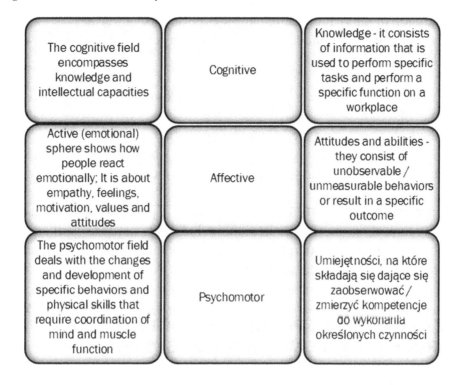

Figure 3. Digital view of Bloom's taxonomy

point is a three-tier model: sender – receiver – communication channels, additionally including the feedback channel absent from the classical approach. A practical exemplification of this relationship shows that the communication model is more complex, since it contains multidirectional relations that can be referred to as chains within which transmitted data, information, knowledge and its aggregates are transformed. We have a multitude of both senders and receivers, with frequently occurring elements that only play a role in the transmission of data, information, knowledge and its aggregates.

The communication model in an organization is shaped by the following groups of factors:

1. The management process structure as the backbone of the communication model;
2. Semantics, meaning the content being sent;
3. The communication system infrastructure (hardware and software), meaning hardware and software tools used in the communication system.

A specific communication system is chosen depending on the situation of the organization under examination, including its needs as well as both tangible and financial resources available. This choice is also influenced by the progress in hardware and software solutions.

THE MANAGEMENT PROCESS STRUCTURE AND ITS ROLE IN THE COMMUNICATION SYSTEM

Management is the element of the communication model that directly affects the communication process structure. This structure is represented graphically or descriptively as an organizational chart. The organizational management chart may be analyzed from the perspective of models of individual (among the various actors in the organization) and global communication (among some or all actors in the organizations and between the organization and its environment). The communication structure is a derivative of the management style of a particular organization. The elements of the communication model include: individual organizational posts, their agglomerates (units, departments, divisions), and existing relationships (ties). The communication model structure both connects and divides in this respect. The basic relationships that are reflected in the communication system are: linear (hierarchical, formal), expressing superior-subordinate relations, and functional, including matrix relationships.

The information and knowledge flow chart illustrated in the communication model is a graphical representation of those relationships. It provides information on: levels of management, the degree of management decentralization or centralization, existing ties, the span of control, to name a few. The communication structure can be defined as a set of elements of a given system with different relations between them. Communication structures can be analyzed and assessed against a variety of criteria. The central criterion is the number of intermediate levels. The number of levels (intermediate tiers) determines both the degree of organizational complexity and the way in which knowledge and information are transmitted and absorbed. There is a tendency to reduce the number of intermediate levels and transmit information and knowledge directly. In many organizations, however, the traditional hierarchical communication system is applied. An exemplary model of organizational units in a hierarchical management system of an organization is shown in Figure 4.

It is believed that an advantage of such a model is that the communication system allows unambiguous and precise transmission of information and knowledge. In practice, nonetheless, the influence of a variety of negative factors prevents the system from performing its tasks. We will return to this problem while presenting semantic and infrastructural aspects of the communication system operation. Attention should also be drawn to the situation described by G. Morgan, namely that organizational hierarchy is a source of different games among employees. When these arise, no substantive issues matter. Workers fight for the position in the organizational structure.

The effectiveness and efficiency of the information system depends on the functioning of its individual tiers as well as on the operation of the so-called feedbacks. Tiers providing appropriate "portions" of information and

Figure 4. Hierarchical management system of an organization

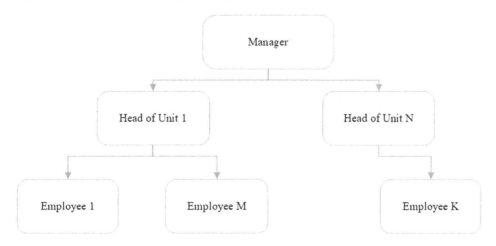

knowledge should get signals of the receipt and understanding of the transmitted content and if the message was an order, of the stage of task completion.

Contemporary communication systems are becoming increasingly network-based and relational.

The network communication system bears the following essential characteristics:

1. Flexible actions undertaken in order to continuously improve the effectiveness and efficiency of the organization as a whole;
2. A constant flow of information and knowledge among all those involved: employees, investors, suppliers, customers, which is connected with cooperation in different cultures and time zones within a uniform global network;
3. Building teams that can quickly deliver high-quality solutions based on the knowledge gained from work in network teams.

In the relational system as presented in Figure 5 and Figure 6, we also have direct links, yet they exist among all employees of the organization whether they cooperate with one another or not. For example, Figure 4 contains no link between employees 2 and K. Links are missing because these workers are not communicating with each other at the moment. The relational system

Figure 5. Model of network communication system

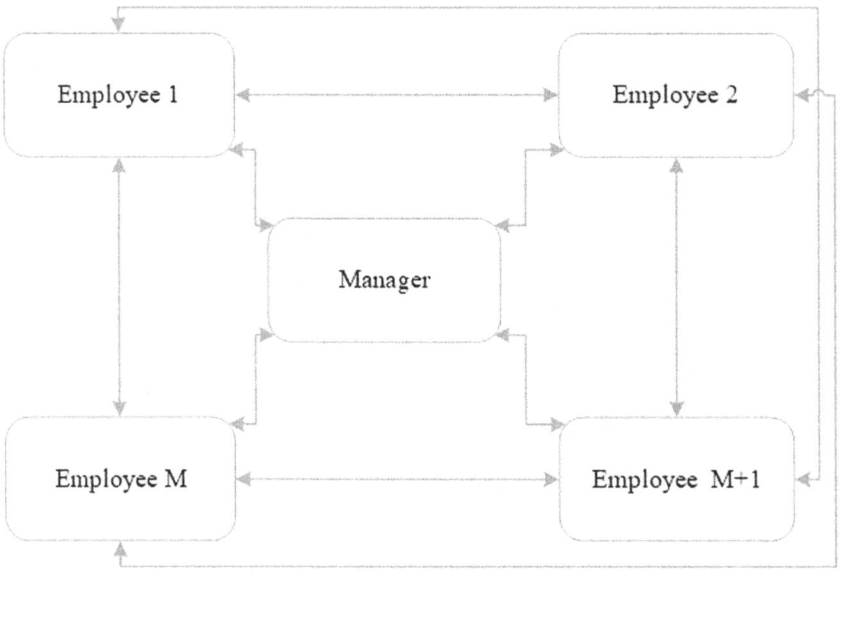

can be accused of being more "lavish" to a certain extent. However, it is more secure than a network system. Among all employees of the organization, there are direct links that can be activated at any time.

Each row of the matrix contains communication links between individual employees in the organization. In practice, a hybrid model is used. It can be referred to as a network-relational or object-relational model where links between objects are network-based and links inside objects are relational. In each analyzed communication system, the role of people responsible for the design of the information and knowledge transmission system also involves building a system of mutual trust between individual members of the organization.

The operation of the network communication system is described later in this chapter. It concerned the implementation of MRP systems and the design and implementation of BI systems in large organizations such as banks, trading and manufacturing companies. The network communication system was well received by the project workers. It turned out to be both efficient and effective. The use of this type of communication made it possible to achieve obvious results such as increased speed of information transfer and a reduced scale of disinformation. What is the added value? This communication model enabled the application of agile methodologies during a research project (Kisielnicki & Misiak, 2016).

SEMANTICS AS A DESCRIPTION OF THE CONTENT TRANSMITTED

A semantic description is the part of a comprehensive communication model that addresses the content and meaning of transmitted data, information, knowledge, and their aggregates. For example, it takes the form of words, phrases, ideas, sentences, and texts. Communication requires more and more information and knowledge that is transmitted in units or as specific aggregates. Global networks such as the Internet make contents accumulate very fast. The development of IT tools today allows large data sets, or Big Data, to be used. According to T. Erl, W. Khattak and P. Buhler (2016), Big Data is a term referring to large, variable and diverse data sets that are difficult to process and analyze; yet, such an analysis is valuable as it can lead to gaining new knowledge. In Laney Douglas' report (Gartner, 2012), Big Data is referred to as a 3V model: a large volume of data; high data velocity; a wide variety of data. It is now assumed that Big Data consists of the following four dimensions, called "4Vs": volume – the amount of data counted in tera- or petabytes; variety – the diversity of data that come from different, often incoherent, sources; velocity – the speed of new data inflow and data analysis, in nearly real time; value – the value of data, the most important data are distinguished from among the mass of insignificant information. Big Data also refers to sets of information that require new forms of processing in order to: support decision-making, discover new phenomena, and optimize processes. Managing large sets is a challenge facing contemporary communication processes.

Apart from sets of information and knowledge, communication systems also comprise networks of relations between individual elements and their aggregates. Sets of notions as aggregates are often termed concepts. Their presentation creates conceptual schemas that, as a description of a certain domain of knowledge, can serve simultaneously as a basis for inference. Such a set of information and relations is defined as ontology. Ontology is a formal representation of a user-defined knowledge domain. It is the records of sets of concepts and relations between them. Creating ontology fulfills the tasks of the users of the communication process in a formalized way (Fensel et al., 2006). Thanks to the communication system, users will obtain the needed data, information, knowledge, ideas, and strategies. Thereby, their information needs are satisfied. The analyzed part of the communication system also includes a set of concepts, techniques and notations aimed at projecting the semantics of data, or their meaning in the outside world. The literature uses different types of notations to present information, knowledge and their aggregates in accordance with the facts (Tatnall, 2010; Dumas et al., 2013; Karaginnis et al., 2016).

How transmitted information and knowledge are recorded can be defined as a semantic model representing a particular ontology. An example of a simple semantic model is the entity-relationship model. The term "semantic model" or "conceptual model" is sometimes also used to refer to a specific diagram (or another linguistic-graphic form) that reflects the reality described by the data. The semantic model consists of a network of concepts and relationships between these concepts. Concepts are ideas, objects or topics of interest to the user.

Effective communication requires a number of factors to be taken into account, regardless of the communication model applied. As written, among others, by L. Beamer and I. Varne (2011), it is important to convey relatively full information and knowledge in this process. Nonetheless, the difficulty of conveying information and knowledge in a multicultural and global world must be realized. Our core task is a knowledge transfer system. It is this system that influences communication models in organizations. M. Polanyi (1974) noticed that there is knowledge that man is not aware of. In his work, he distinguished two types of knowledge:

1. Tacit knowledge existing only in the mind of the person who possesses it, produced by experience and not fully realized ("I know that I can do it"), manifested only through skillful action,
2. Explicit knowledge (formal knowledge) expressed as signs and recorded on knowledge carriers.

The word "tacit" derives from the Latin term *tacitum* meaning secret, hidden, unrevealed. In the communication system, we should pay particular attention to cultural problems that can be categorized as tacit knowledge. The weight of this issue is noted by many researchers. The works by Ch. Hampden-Turner and A. Trompenaars (1997; 2003; 2017) report interesting research in this respect. Inter-cultural communication is a meaningful part of communication processes that affect communication models in organizations. According to F. Trompenaars and Ch. Hampden-Turner (2002), the condition of the economy is not determined solely by economic laws. Also communication systems influence the economic situation of organizations and countries. Their research covered a group of 15 thousand senior managers from 43 different countries who were responsible for international projects. The results showed that even people who performed international tasks and were in constant contact with representatives of other nationalities used national stereotypes in communication systems. Moreover, what could be noticed was that those stereotypes were reinforced in their case. Those works resulted in so-called Trompenaars' model. It is a framework for semantic analysis of inter-cultural communication and serves to understand communication systems and activities of international corporations. It can be employed to design communication systems and to understand the content being transmitted. There are seven components of the model:

1. Universalism vs particularism (What is more important, rules or relationships?);
2. Individualism vs collectivism (communitarianism) (Can we function in a group or individually?);
3. Neutral vs emotional (Do we display our emotions?);

4. Specific vs diffuse (How to separate our private and working lives?);
5. Achievement vs ascription (Do we have to prove ourselves to achieve a status or is it given to us?);
6. Sequential vs synchronic (Do we do one thing at a time or several things at a time?);
7. Internal vs external control (Are we able to control our environment or are we controlled by it?).

Understanding these factors can help us control our behavior and know and understand the behavior of people who grew up in other cultures.

The analysis of research results is based on the view that the communication system, including the knowledge transfer system, should comprise, among others, solutions that take into account factors such as values, habits, and applied cultural models. How communication process participants perceive time, family, history, ethics in progressing the career ladder and how they see the hierarchy of needs – all this, even subconsciously, influences decision-making processes. The diversity of the world we live in should be taken into account. We should remember that there are differences in adopted management styles and conflict-resolution and negotiation tactics in other countries. If we not only remember but also know them, we can effectively communicate in a multicultural world. It should always be borne in mind that a prominent element of information is the context of what we communicate. It is not always possible to send the context directly, yet – as presented later in this monograph – software development should allow this inconvenience to be removed in the near future.

COMMUNICATION SYSTEM INFRASTRUCTURE

The use of data, knowledge and information resources and their transmission requires infrastructure appropriate to the management system and to the semantic content of information and knowledge. Infrastructure consists of hardware and software tools as well as networks connecting them. At this point, we wish to focus on some elements that are essential to the communication process. Their characteristics are outlined in many publications. The most recent and comprehensive specification is presented in the previously mentioned ten-volume encyclopedia edited by M. Khosrow-Pour (2017). The changes in the infrastructure supporting communication systems are directed towards building faster and faster computers with increased capacities. It should be taken into account that new possibilities in the communication system will occur when quantum computers become widespread. The reason will be that the quantum bit (qubit) does not have a fixed value of 1 or 0 like a standard computer, but may remain in an intermediate state. The application of this notation will have a significant impact on the efficiency of computers as tools supporting complex decision-making based on multiple criteria and will help solve multidimensional decision problems.

The establishment of network organizations is aided by the emergence and development of mobile technologies. It is their development that contributes to the enhanced computing power of mobile devices, increased data throughput, and users gaining access to databases and knowledge whenever and wherever they need them. Mobile technologies contribute substantially to the changes in both business and society. They have led to the formulation of new rules and patterns of communication behavior. In their works, researchers clearly pointed to the immense potential of wireless solutions for the development of communication systems (Wan et al, 2013). Even the most complex management structures can work in this way. The intensive advancement of mobile technologies in recent years has enabled the improvement of communication processes. This has been achieved through: deployment of mobile infrastructure, smartphones, and accelerated data transmission processes. With the rapid development of mobile technologies, it will soon be possible to talk about organizations whose employees have access to the data they need, regardless of place and time. As a result, mobile technologies allow for the minimization of constraints ensuing from the distance between the sender and the receiver. Research conducted by international teams such as Chen Xu, lingyang Song, Zhu Han, Qun Zhao, Xiaoli Wang, Xiang Cheng and Bingli Jiao (2013) or Wendi L. Adai, Nancy R. Buchan, Xiao-Ping Chen, Michael G. Foster and D. Liu (2016) on models called: Device-to-Device (D2D) or Peer-to-Peer (P2P) (Figure 6) show the change of the role of the individual elements of the computer network. These changes are aimed at supporting the management of network organizations. The contemporary communication model within a computer network provides all hosts[1] with the same authorizations. This is interchangeability depending on the specific client-server architecture.

As argued by M. Castells, mobile solutions result in the emergence of a new power theory based on the management of communications networks in the information era. Hence, we can recognize that hierarchical structures

Figure 6. Client-Server vs P2P model

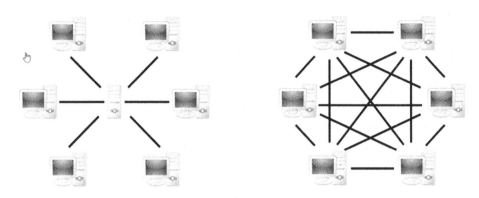

typical of traditional organizations are gradually making room for network structures. Network organizations use BI systems, in particular Business Activity Monitoring (BAM) solutions, to a greater extent than traditional ones. Their application allows the operation of the entire organization to be analyzed comprehensively in real time (Gartner, 2013; Baumann et al., 2015)

Computer-aided or computer-assisted is an adjectival phrase that hints at the use of a computer as an indispensable tool in a certain field, usually derived from more traditional fields of science and engineering. Instead of the phrase computer-aided or computer-assisted, in some cases the suffix management system is used.

Contents:

1. Engineering and production
2. Music and arts
3. Human languages
4. Medicine
5. Software engineering
6. Traffic control
7. Teaching
8. Mathematics
9. Economy
10. Communications
11. Security
12. Entertainment

Business Activity Monitoring (BAM) includes software that analyzes processes and technologies so that, throughout a network organization or its part, it is possible to analyze key business performance indicators based on real-time data. Such solutions increase the operational speed and efficiency of a network organization. Hence, it is possible to reduce some inconvenience involved in the functioning of a network organization. Network organizations can be alleged to be less flexible and not to respond quickly enough to emerging crisis threats as compared to traditional organizations. The BAM model is composed of three parts, namely the modeling process, performance indicators, and information dashboards. Such dashboards allow organizations to obtain quick information about what is happening inside. They are tools that enable managers to respond to emerging threats, which is consistent with the Real Time Management concept elaborated on in Chapter 3 of this monograph.

When comparing communication infrastructures of network and traditional organizations, we can assume that any hardware and software solution can support the management system of a hierarchical organization. This is not the opposite relationship. A network organization, even with a decentralized management system, needs modern infrastructure with high network capacity as well as Big Data technology supported by Business Intelligence systems together with tools such as Business Activity Monitoring. As claimed by Elaine B. Kerr and Starr Roxanne Hiltz (2013), international network organizations with inter-cultural management are much more demanding than national network organizations.

The infrastructure in place has a major impact on the type of communication system used. Traditional and network organizations adopt both online and offline systems. This distinction stems from the fact that in order for the communication system to be considered an online one, it must be available for immediate use on demand without human intervention. The communication system is under direct control of another computer communication system. On the other hand, an offline communication system does not meet this condition. Its main power source has no direct contact with the computer communication system (it is disconnected or turned off). The distinction between the online and offline mode is traditionally recognized as the distinction between computer communication and face-to-face contact (e.g. negotiation or direct arrangements). Online means virtuality or cyberspace, whereas offline is reality. Following the case studies presented in the works by P. R. Smith, Ze Zook, (2017), K. Subrahmanyam, S. Reich, N. Waechter,

G. Espinoza (2008), a thesis can be put forward that network organizations, notably those active in marketing, logistics, services and social media, use online systems. D. Slater (2002) states that this distinction is simplistic. In order to support the argument that the distinctions in relations are more complex than the simple online vs. offline dichotomy, he notes that some people do not distinguish between online and offline relations. He claims that even a telephone conversation can be regarded as an online experience in certain circumstances and that the distinction between the applications of different technologies (such as PDA vs. mobile phone, online television vs. Internet, telephone conversation vs. Voice over Internet Protocol) is blurred, which has made it impossible to use the term "online" in the meaning adopted by the first generation of online research. He also speculated that the difference between online and offline might be seen as rather peculiar and incomprehensible within 10 years to come. However, it is 2018 now, and D. Slater's forecasts have not become a reality. Yet, K. Melis, K. Campo, E. Breugelmans, L. Lamey (2015) are of similar opinion. They believe that the difficulty in distinguishing between online and offline communication systems will arise in the consumer-vendor system. The basic determinants of the specific "switch" between channels include the propensity of buyers to choose channels regardless of their type (channels perceived by them as the most suitable in the context of a particular purchase), low cost of channel switching, and inter-channel synergy ensuring that buyers using multiple channels are satisfied with their use no matter where they are located.

According to the Berg Insight report (Reports from Berg Insight, 2018), the combination of online and offline communication systems results in the development of further economic sectors. As assessed by that research firm, telematic systems, which combine telecommunications, IT and informational infrastructural solutions, will significantly grow in popularity in the coming years. Currently, telematics is applied in many economic sectors. Already a few years ago, it was appreciated by the leaders of the transport and logistics industries. Telematic services collect accurate data about the movement of fleets. Thereby, truck owners can track business activity in real time and precisely estimate, for example, which stores need more goods and in which quantities. There are products available in the market that allow the monitoring of motor vehicles (TIR) and tank vehicles by means of the GPS system. It may be presumed that Berg Insight's considerations concern network organizations.

DETERMINANTS OF COMMUNICATION IN NETWORK ORGANIZATIONS

The effectiveness of communication systems in network organizations depends on a wide range of factors. The impact of determinants on the improvement project is not identical. We can divide them into:

1. **Absolute:** We cannot spend more on the improvement of the organization's communication system than a certain amount, we cannot install only a piece of software or hardware we need;
2. **Relative:** Which can be overcome with a particular effort. For example, installing the software that we consider necessary requires the involvement or training of staff or the installation of new types of computers, or may call for some adaptation of the premises;
3. **Apparent:** Which appear to be significant, but on closer examination, it turns out that removing them requires little effort. For example, we want to introduce a new type of software, and it seems that we will have difficulty training staff, yet – on a careful analysis of IT staff skills – it turns out that it is the software that employees know but have never had the opportunity to use it.

The determinants can be grouped into the following categories:

1. Technical, related to technical and technological progress in the area of IT infrastructure, i.e. hardware, both core and application software, and the communications system (attention should be paid to access to global networks such as the Internet and the speed of information transfer);
2. Economic, referring to both the amount of financial resources and formulas of profitability for innovative activities measured by various evaluation indicators of output, financial incentives;
3. Organizational, depending on the applied management formula (centralization-decentralization), organizational structure, application of outsourcing solutions;
4. Sociological and psychological, i.e. preferences, management styles, negotiation and conflict-resolution methods, organizational culture, non-financial incentive systems, ethical issues;
5. Legal, existing legal regulations, i.e. certificates, quality management system, safety standards.

Among economic determinants, financial factors play a special role. The organization allocates certain resources for the modernization of its communication system. They can be used both for hardware and software purchases and for the improvement of qualifications of system users. Practically, the money for modernization is a specific amount limited by two thresholds. The "upper" threshold means that we cannot spend more money than we have. The "bottom" threshold means that no solutions to improve the communication system may be introduced until a certain minimum amount of funding needed to carry out an innovative project is allocated. The "upper" one implies that we cannot spend more funds than we have because such overspending would have a negative impact on the functioning of the organization.

The strength of each determinant is different. In analyzing the influence of individual factors, a simulation calculus is helpful as it answers the question: What do we gain if the constraint is overcome? The simulation approach requires action and reflection on whether the money spent will be used optimally. As demonstrated by the analyses performed, it is the technical barrier that most affects users. Modernization of the communication system necessitates the use of appropriate IT equipment. What can determine communication system modernization is the computer itself, mobile devices or peripherals such as optical readers for large data volumes, scanners, printers, and data transmission devices. A technical determinant is also a lack of access to necessary computer systems, especially those connected with knowledge acquisition and management systems.

Economic determinants pertain to situations where we cannot implement the most needed information system due to a lack of sufficient financial resources. Modernization of the communication system is influenced by prices and taxes. An organization that does not have adequate resources to buy hardware chooses to lease it or uses cloud computing, as highlighted by Qi Zhang, Lu Cheng, Raouf Boutaba. However, organizations sometimes opt for the cloud technology not because of expenditure but because this solution is more convenient, safer, etc., for them at the moment. Thus, the user can upgrade the communication infrastructure with less resources than when buying necessary hardware.

Organizational determinants are equally common in business practice. There are objections against many communication systems because of the problems with coherence of diverse infrastructure. Integrators and inter-operable platforms do not always work properly. The location of the team of system administrator(s) managing the entire communication is a particularly sensitive issue. How the administrator is linked with all organizational units is not always clear. Not fully defined competences and responsibility are still another element classified among organizational determinants.

The socio-psychological barrier is a natural reaction to changes in the organization. It can be termed immunological barrier or negative coopetition, as indicated by Paolo Parigi, Jessica J. Santana and Karen S. Cook (2017). Members of the organization where changes are forthcoming try to discredit the changes and it is on this that they focus their activities. In extreme cases, manipulation of the transmitted information may undermine confidence in the communication system. A mechanism may also be triggered whereby information unfavorable to the sender is not conveyed. The opinion survey of designers in MBA studies that I conducted in 2014–2016 on the applicative efficiency of IT systems allowed the elements negatively influencing communication systems in organizations to be specified. The students most often referred to the following as communication barriers within and between organizations:

1. Poor quality of the source data, including their incomparability resulting from different rules for their acquisition;
2. Inadequate hardware and software, including financial problems, meaning a lack of funds for appropriate equipment, and a lack of user qualifications for using the systems;
3. Unspecified requirements both for the speed of the system operation and for the scope of needed information;
4. Unclear definition of economic and legal rules for the communication system operation, including unspecified roles and authorizations of individual participants in communication processes;
5. Routine and resistance to new solutions as well as reluctance to transmit information and knowledge that senders deem unfavorable to themselves;
6. Long-standing or imaginary dislike for the receiver – here, the reasons may be very different, often cultural and frequently irrational (I do not like inhabitants of city X and I will not help its inhabitant).

The studies by T. Ariyachandra and H. Watson (2006) focused on three groups of determinants:

1. **Organizational Factors:** Are planned changes in the communication system supported by the management board and the sponsor and are the objectives and vision of the solution clear?

2. **Factors Related to the Efficiency of the Communication Process:** Is it oriented towards business needs and interactive change implementation and management?

3. **Factors Related to Infrastructure Capacity (in Terms of Technology):** Are factors such as the system design focused on business and the user, system scalability, data quality and integration appropriate?

W. Delone and E. McLean (2003) identified the following determinants of today's communication systems: quality of information, quality of the system (including its functionality and ease of use), quality of the system maintenance. These factors directly translate into the requirements for the use of communication systems and the satisfaction of system users. Identifying, and then overcoming, the determinants affects the operational efficiency of any communication system.

COMMUNICATION IN NETWORK AND HIERARCHICAL ORGANIZATIONS AND ITS IMPACT ON ICT PROJECT SUCCESS: RESEARCH RESULTS

The analysis of the research evaluating successes and failures of IT projects was performed on the basis of the ISBSG database. The database contains historical data on completed IT initiatives from a number of public and private institutions operating in various industries and administrations. These initiatives concerned both development of new and modification of existing software. The analysis shows that most IT projects exceed the pre-defined cost and time. Many publications offer a more detailed examination indicating the reasons for IT project failures. L. Mieritz (2012) and S.J. Spalek (2005) list the following reasons for failures: lack of communication within the team and with stakeholders, unrealistic budget and schedule, complexity and a prototypic nature of IT systems, lack of sufficient resources or team qualifications.

I have been conducting comparative analyses of the quality of communication systems within R&D project teams for many years. Here, I wish to present their synthetic results. The reference list contains monographs that outline the analyses of individual projects (Kisielnicki, 2013; 2017). Due to a relatively long time span of the research, detailed comparisons could not be made between individual projects. The factors distorting such examination included: sizes of projects and their different scopes, differences in the infrastructure used, differences in costs and times of implementation. However, expert opinions on the quality of communication systems in project management, that is on successful implementation, may be said to provide quite precise assessments. It is assumed that project success is decisively influenced by the management system that, in turn, depends on the quality of the communication system as a system for transmitting information and knowledge.

The research focused on comparing two most popular communication systems used in project teams: traditional (hierarchical) and network. Both models were presented in Figures 4 and 5 above. The research on communication systems in project teams was done for projects carried out in different years (1996–2015). The investigation covered 28 management systems for R&D projects concerning the design or implementation of IT systems. I participated personally in 18 projects as a performer of tasks or a project manager. The data for research on other projects were obtained from documentation and interviews with project participants. My involvement in those projects concerned supervision over their implementation[2]. Difficulties in conducting the research were caused by the practical impossibility to compare the effects produced by two teams using different communication systems, due to the unique nature of the analyzed situations (which is a rule in research projects). Such an experiment would require identically qualified teams carrying out the same project, hence the quantitative results are based only on expert estimates.

Most studied projects involved improvement of the existing business management system. Substantial materials were collected during the implementation of projects covering, among others: IT application in accounting for big textile mills, improvement of the IT-supported benefit calculation system in the Polish Social Insurance Institution, implementation of the MRP II/ERP system in a pharmaceutical company, IT applications to support the management system for a regional capital city, development of an IT application strategy for the National Bank of Poland (NBP), IT application to improve the management system for prisons and the police, participation in the SYNAT project (development of a national scientific research information system), design of a BI system supporting organizational creativity.

This set of projects provides a picture of a very broad research spectrum. The investigation covered both business (20 projects) and administrative (8 projects) organizations. Although most projects (19) were successful, they had encountered many threats in the course of implementation. The success was assumed to be a situation where the planned project scope, implementation time and costs had not increased by more than 10% in the previous two years.

Teams of 20–60 people took part in the implementation of each project. These were both designers and cooperating specialists in various industries. Such a minimal size of the implementing team was chosen because I was interested in teams where implementation required cooperation of various teams. The research questions that I tried to answer were:

1.   What is the effectiveness of basic communication systems in project teams (do the team members feel that they received full information and knowledge about the performed tasks from other team members)?
2.   What rules should be applied to manage a project team so as to ensure the best flow of information and knowledge in achieving the set objectives?
3.   What communication system can be recommended for the implementation of R&D projects?

The effectiveness of a communication system is understood as the degree to which the project objectives have been attained, i.e. the success as specified above. The analysis covered: documents (conceptual design and documentation of project implementation including: costs, times, scopes), special questionnaires filled in by both selected managers and designers, reports on the implementing team's discussions (in which I participated) about the causes of failures or difficulties in project implementation.

Discussions were the primary research method, supplemented by the analysis of documents and questionnaires. Documents and questionnaires showed that there were problems to be solved. Discussions about changes in the management system provided recommendations on what should be done to address these issues. Experts were very often involved in the discussion on the causes of threats. They were outsiders including specialists from cooperating teams as well as users. Any major deviation from the assumed standard (plan, schedule) was debated and considered. The analysis of documents and questionnaires verified the answer to the question of whether the decisions taken had been effective. One of the basic questions asked after project completion was: Would you like to work on the next project in the same team?

In order to enhance the analysis of the communication system, the examination occasionally covered the time within which the information given to a particular person would reach all team members. In that case, the information was transmitted as an e-mail, and the time when the message would be read was analyzed. The frequency of using a database or the frequency of sending and receiving e-mails was also investigated. The ancillary question concerned the degree of understanding and usefulness of information to the recipient. The combination of the speed of information transmission, information comprehension, usefulness and appropriate use has a decisive influence on project success understood as its implementation within the planned time span, scope and costs.

The factors that disturb communication processes in project implementation can be divided into the following groups:

1. **Caused by External Factors:** Delay in the delivery of technical and financial resources, inadequate documentation provided by the user, change of legislation, change in the management system of the organization for which the project is carried out, random factor, and a team member's incapacity for work (illness);

2. **Resulting From Internal Factors Such as:** Poor communication, lack of knowledge and experience in project implementation, conflicts between team members, errors in project management.

All the factors listed as internal are very strongly linked with the communication system. The research results formed the basis for improving the existing information transmission system. Communication systems were also analyzed against the following criteria:

1. Deviations from planned costs, time, adopted parameters;
2. Failure to spot threats in time;
3. Conflicts during task performance;
4. Lack of cooperation and failure to share knowledge with partners;
5. Willingness to work on the next project in the same team.

The examination of those communication systems revealed that:

Out of 18 network communication systems, 15 projects had been successful (i.e. over 80%) according to the previously adopted criterion;

Out of 10 traditional communication systems, 5 projects had been successful (i.e. 50%).

Certainly, not only communication systems are responsible for success. However, the answer to the previously mentioned question: "Would you like to work again in the same team?" was symptomatic. The following responses were given in this respect:

1. **Traditional – Hierarchical Communication System:** Positive responses: 60–70% of managers and about 30% of designers.
2. **Network Communication System:** Positive responses: 70–80% of those surveyed (it was impossible to make a precise differentiation between managers and designers for this system).

The number and gravity of conflicts were much lower in teams working in network communication systems than in hierarchical systems. The analysis of the results showed that data transmission time was about 30% shorter in network systems than in hierarchical ones.

The research on the functioning and effectiveness of communication systems make it possible to argue that the network system has an advantage over the traditional (hierarchical) one, notably as regards the following elements:

1. **Monitoring of Implementation:** Risks to implementation and deviations from planned costs, time, adopted parameters of IT systems were spotted earlier than in hierarchical systems, hence intervention decisions could be made in good time. Also, almost all staff working in network systems felt responsible for the project;
2. **Cooperation and Knowledge Transfer in Task Performance:** There was good cooperation between co-workers in task performance and transfer of information and knowledge to their partners. No artificial barriers existed such as leaders and employees. Generally, each worker is a leader and a performer of tasks, depending on the situation;
3. **Problem Solving:** Conflicts in task implementation were much less severe than in hierarchical systems, and if they arose, they were quickly resolved within the team.

The communication system used in a network team requires a number of conditions to be met. The level of qualifications of individual employees and their willingness to work together are most important. Such a system is difficult for individualists and people wishing to make an administrative career. In the recommended system, the career path is directly associated with professional development. Nonetheless, it should be realized that this is a system where projects are difficult to manage. The leader of the entire project bears a great responsibility, being in charge of team selection and organization and creation of an appropriate climate. Compared to the hierarchical system, the project leader should give up many managerial powers and delegate them to the implementing teams. Yet, the leader's responsibility for project implementation does not change. Therefore, many project managers, even realizing its limitations, prefer the hierarchical system as allowing for easier control of the work done by their subordinates and strengthening their formal authority.

It is quite difficult to provide unequivocal answers to the questions:

1. Which of the analyzed communication systems is more effective under all circumstances of research project implementation?
2. What is the economic efficiency of replacing a hierarchical communication system with a network one?

Effectiveness and efficiency depend on manifold factors. The research by A. Dulbiński found that about 20% of the surveyed employees in various business areas were concerned about network communication systems. The respondents believed that such communication systems did not ensure adequate security of transmitted information. For that reason, I would like to conclude by noting that it is not only the communication system that determines the working efficiency of the project team. It is also very important to select project team members and decide on the incentive system for those implementing the project. The former problem has already been flagged up, but the general principle should be to support communication systems through appropriate incentives. Developing an incentive system is a separate issue strongly dependent on the organizational culture and conditions prevailing in the labor market. The system will be different in India, Great Britain, Poland and the United States. Nonetheless, the communication system is the element that influences the efficiency of the project team regardless of the project scope and type, the method of employee selection or incentives offered.

CONCLUSION

Effectiveness and efficiency of a communication system depend on manifold factors. It is not always possible to identify them. Therefore, we wish to point out that not only communication models determine the efficiency of information and knowledge transmission systems. Some factors influencing this process are highlighted at the end of this chapter. The work on this issue should be continued. However, as described in the article, the operation of communication systems is highly dependent on communication models employed. In most situations, the network model has an advantage over the hierarchical one. This is evidenced by our research and the related literature alike.

REFERENCES

Adai, W. L., & Buchan, N. R. (2016). A Model of Communication Context and Measure of Context Dependence. *Academy of Management Discoveries*, 2(2), 198–217. doi:10.5465/amd.2014.0018

Anderson, L. W., & Krathwohl, D. (2001). *A Taxonomy for Learning, Teaching and Assessing: a Revision of Bloom's Taxonomy of Educational Objectives*. New York: Longman.

Ariyachandra, T., & Watson, H. (2006). Which Data Warehouse Architecture Is Most Successful? *Business Intelligence Journal*, 11(1).

Baumann, F., El Hussein, R., & Roller, D. (2015). State of the Art of BPM - Approach to Business Process Models and its Perspective. *International Journal of Electronics Communication and Computer Engineering*, 6(6), 2278–4209.

Beamer, L., & Varne, I. (2011). *Intercultural communication in the global workplace*. Boston: McGraw-Hill/Irwin.

Berg Insight. (2018). *The Global M2M/IoT Communiactions Market – Report*. Retrieved from: https://www.marketresearch.com/Berg-Insight-v2702

Bloom, B. S., & Engelhart, M. D. (1956). *Taxonomy of educational objectives: The classification of educational goals. In Handbook I: Cognitive domain*. New York: David McKay Company.

Castells, M. (2013). *Communication Power*. OUP Oxford.

Delone, W., & McLean, E. (2003). The DeLone and McLean Model of Information Systems Success: A Ten-Year Update. *Journal of Management Information Systems*, 19(4).

Dulbiński, A. (2012). *Doskonalenie procesu zarządzania przedsiębiorstwem z wykorzystaniem sieci rozległych (praca doktorska)*. Warszawa: Wydział Zarządzania Uniwersytetu Warszawskiego.

Dumas, M., La Rosa, M., Mendling, J., & Reijers, H. A. (2013). *Fundamentals of Business Process Management*. Springer. doi:10.1007/978-3-642-33143-5

Eadie, W. F., & Goret, R. (2013). Theories an models of communication: Foundations and heritage. In Theories and models of communication (Tomo i). Berlin: De Gruyter Moution. doi:10.1515/9783110240450.17

Erl, T., Khattak, W., & Buhler, P. (2016). *Big Data Fundamentals: Concepts, Drivers & Techniques*. Prentice Hall Press Upper Saddle River.

Fensel, D., & Lausen, H. (2006). *Enabling Semantic Web Services: Web Service Modeling Ontology*. Springer.

Gartner. (2013). *Business activity monitoring (BAM)*. Retrieved from: http://www.gartner.com/it-glossary/bam-businessactivity-monitoring

Karagiannis, D., Mayer, H. C., & Mylopoulos, J. (Eds.). (2016). *Domain-Specific Conceptual Mod-eling*. Springer. doi:10.1007/978-3-319-39417-6

Kerr, E. B., & Hiltz, S. R. (2013). *Computer-Mediated Communication Systems: Status and Evaluation*. Academic Press.

Khosrow-Pour, M. (Ed.). (2017). Encyclopedia of Information Science and Technology. Hershey, PA: IGI Global.

Kisielnicki, J. (1994). *Informatyczna infrastruktura zarządzania*. Warszawa: Wydawnictwo Naukowe PWN.

Kisielnicki, J. (2006). Transfer of Information and Knowledge in the Project Management. In E. Coakes & S. Clarke (Eds.), *Encyclopedia of Communities of Practice in Information and Knowledge Management* (pp. 544–551). London: IDEA Group Reference. doi:10.4018/978-1-59140-556-6.ch091

Kisielnicki, J. (2008). *Intellectual capital In the knowledge management process – relations- factors w pracy zbiorowej pod redakcją A.Z. Nowak, B. Glinka, P. Hensel, Business Environment in Poland*. Wydawnictwo W.Z. UW.

Kisielnicki, J. (2014a). *Zarządzanie projektami badawczo-rozwojowymi*. Warszawa: Wolters Kluwer.

Kisielnicki, J. (2014b). *Zarządzanie i informatyka*. Warszawa: Placet.

Gartner. (2012). *Deja VVVu: Others Claiming Gartner's Construct for Big Data*. Author.

Melis, K., Campo, K., Breugelmans, E., & Lamey, L. (2015). The impact of the multi-channel retail mix on online store choice: Does online experience matter? *Journal of Retailing*. Retrieved from: https://lirias.kuleuven.be/bitstream/123456789/472387/1/2014-12-04+-+Manuscript_.pdf-

Mieritz, L. (2012). Survey Shows Why Projects Fail. Gartner Research Report.

Morgan, G. (2013). *Obrazy organizacji*. Warszawa: PWN.

Mullins, L. (1993). *Management and Organizational Behavior*. London: Pitman Publishing.

Parigi, P., Santana, J. J., & Cook, K. S. (2017). Online Field Experiments, Studying Social Interactions, *Context. Social Psychology Quarterly*, *80*(1), 1–19. doi:10.1177/0190272516680842

Polanyi, M. (1974). *Personal Knowledge: Towards a Post-Critical Philosophy*. University of Chicago Press.

Qi, Z., Cheng, L., & Boutaba, R. (2010). Cloud computing: State-of-the-art and research challenges. *Journal of Internet Services and Applications*, *1*(1), 7–18. doi:10.100713174-010-0007-6

Schahaf, P. (2008). Cultural Diversity and Information and Communication Technology Impacts on Global Virtual Teams: An Exploratory Study. *Information & Management*, *45*(2), 131–142. doi:10.1016/j.im.2007.12.003

Shannon, C., & Weaver, W. (1948). *A Mathematical theory of communication*. University of Illinois Press.

Slater, D. (2002). Social Relationships and Identity On-line and Off-line. In L. Lievrouw & S. Livingstone (Eds.), *Handbook of New Media: Social Shaping and Consequences of ICTs* (pp. 533–543). Sage Publications Inc. doi:10.4135/9781848608245.n38

Smith, P. R., & Zook, Z. (2017). *Marketing Communications: Offline and Online Integration, Engagement and Analytics*. London: Kogan Page LTD.

Spalek, S. J. (2005). Critical Success Factors in Project Management – To Fail or Not To Fail, That is the Question! *PMI Global Congress Proceedings.*

Stawnicza, O., & Kurbel, K. (2012). How to Prevent before You Must Cure - A Comprehensive Literature Review on Conflict Management Strategies in Global Project Teams. *Proceedings of the International Research Workshop on IT Project Management.*

Subrahmanyam, K., Reich, S., Waechter, N., & Espinoza, G. (2008). Online and offline social networks: Use of social networking sites by emerging adults. *Journal of Applied Developmental Psychology*, *29*(6), 420–433. doi:10.1016/j.appdev.2008.07.003

Tatnall, A. (2010). *Web Technologies: Concepts, Methodologies, Tools and Applications*. Hershey, PA: Information Science Reference. doi:10.4018/978-1-60566-982-3

Trompenaars, B. F., & Hampden-Turner, Ch. (1997). *Riding the Waves of Culture: Understanding Cultural Diversity in Business* (2nd ed.). London: Nicholas Brealey Publishing Ltd.

Trompenaars, B. F., & Hampden-Turner, Ch. (2002). *Siedem wymiarów kultury. Znaczenie różnic kulturowych w działalności gospodarczej*. Kraków: Oficyna Ekonomiczna.

Trompenaars, B. F., & Woolliams, P. (2003). *Business Across Cultures*. Capstone Publishing Ltd.

Trompenaars B. F. Hampden-Turner Ch. (2017). *Culture for Business*. Retrieved from: http://www2.thtconsulting.com

Wan, J., & Zhuohua, L. (2013). Mobile Cloud Computing: Application scenarios and service models. *9th International Wireless Communications and Mobile Computing Conference (IWCMC)*, 644 – 648.

Xu, Ch., Song, L., Han, Z., Zhao, Q., Wang, X., Cheng, X., & Jiao, B. (2013). Efficiency Resource Allocation for Device-to-Device Underlay Communication Systems: A Reverse Iterative Combinatorial Auction Based Approach. *IEEE Journal on Selected Areas in Communications*, *31*(9), 348–358. doi:10.1109/JSAC.2013.SUP.0513031

Woods, D., & Hollnagel, E. (2005). *Joint Cognitive Systems: Foundations of Cognitive Systems Engineering*. Boca Raton, FL: Taylor & Francis.

ENDNOTES

1. Host – any machine (computer, network adapter, modem, etc.) involved in data exchange or providing network services through a computer network by means of a communication protocol.

2. For many years, I was the CEO of one of the biggest Polish software houses (Centrum Projektowania i Zastosowań IT ZETO ZOWAR). I was responsible for the Project and Research Department.

Chapter 5
Decision–Making Systems in Traditional and Network Organizations

ABSTRACT

Decision-making processes taking place in increasingly complex traditional and network organizations require the use of modern decision support systems. As a result of these solutions, decisions are made to support the development of the organization, its modernization, and thereby lead to increased competitiveness. The subject of the analysis of decision-making systems in organizations has been explored in a number of publications. This chapter addresses selected problems concerning the design and functioning of the decision-making system in traditional and network organizations. Particular attention was paid to the analysis of the decision-making process and the tools used to support this process. The results of research on evaluation of available solutions, especially in the field of information technology, in decision-making processes in network organizations were also presented.

DECISIONS IN ORGANIZATIONS. DETERMINANTS OF DECISION-MAKING PROCESSES

A decision is a deliberate, non-random choice of one possible solution to a particular problem. It is a conscious act of the will of a person or a team that makes it. Decisions can be made when the following conditions are met:

DOI: 10.4018/978-1-5225-5930-6.ch005

1. The decision is the result of a choice in which the condition for decision-making is the existence of at least two possible solutions.
2. The decision is closely related to the person or persons who make it, and who is/are capable of making it both in terms of their emotions and skills.

Decisions can be analyzed narrowly (results) and broadly (actions). The narrow approach is understood in this case as the final result of the decision, i.e. the choice that is made by the decision-maker in a conscious and deliberate way. In action terms, a decision is the process of deciding, or how a particular decision is made. The decision-making process consists of steps such as defining the problem, setting goals, identifying resources and constraints, identifying possible solutions, selecting one solution, implementing it, and evaluating the outcome.

Regardless of the nature of an organization - traditional or network - the following elements can be identified in the process (Kisielnicki, 2012, 16):

1. **Decision Situation (Problem):** When a decision maker chooses one of at least two possible solutions, a variation of action; a decision-making situation is the set of conditions that influence a decision;
2. **Decision-Maker:** Person or, in the case of a collective decision, a group of persons who choose the solution;
3. **Reason for the Decision:** Impulse (e.g. an opportunity or a threat), situation that requires making a decision;
4. **Purpose of the Decision:** Target status achievable as a result of the implementation of the decision;
5. **Subject Matter of the Decision:** What the decision concerns;
6. **Decision:** Outcome of the decision-making process;
7. **Decision-Executor:** Actor, not always decision maker, who executes the decision;
8. **Decision-User:** Person for whom the decision may be relevant.

A decision is always related to the concept of change. Such change may be the decision to transform the organization, e.g. from a traditional into a network one, or the development of a network organization. An important premise to be accounted for during the process is the issue of responsibility for the outcomes and consequences of decisions, which imply not only economic but also social and ethical effects (Walczak, 2012). The consequences of decisions can vary from internal (i.e. affecting the organization, its members, or financial performance) to those of the inside-out effect. In the latter case, there is an entire range of consequences that affect the environment. Impact on the environment, understood in a very broad way (i.e. not limited only to customers and competitors), is a problem that is being increasingly analyzed both in science as well as in the practical management of organizations. This topic is defined collectively as *corporate social responsibility* (CSR) and it is understood as a strategic opportunity and a set of commitments that the organization should analyze in its decision-making processes (Davis, 2005, 87). There are dozens of definitions of Corporate Social Responsibility. The definition that is officially accepted by United Nations is "a management concept whereby companies integrate social and environmental concern in their business" (UNIDO, 2017). The main goal of all CRS actions are: to control impact of business for society and to increase benefits provided by business to the community and to fulfill the main principle "do not harm" any more.

Another assumption in analyzing the decision-making process is that the decision maker behaves in a logical and rational manner, and the decision is fully in line with the objectives of the organization. The basis for the classical approach to decision-making is the conviction of making an optimal (balanced) decision, which is easier to do in traditional organizations.

Decision-making processes in traditional organizations are supported by a number of IT solutions mentioned in Chapter 3 of this monograph, which include integrated systems supporting business activities such as MRP, ERP and BI. They may also be sectoral systems that are specific to the particular industry in which the company operates. As an example, GDS Amadeus can be used as an example of software that supports the aviation and tourism industry, or BIM (Building Information Modeling) solutions that are opted for in the design and implementation of building investments. The situation gets more complex for network organizations, which not only are interconnected through intangible resources (as often referred to in the content of the study), but also through tangible ones. As a result, network projects are implemented by technical solutions that remain at the disposal of each participant in the project. Consequently, there may be many often incompatible solutions and tools. In this case, lack of cooperation will stem from technical (e.g. different formats of stored data causing incompatibility), personal (e.g. unwillingness of the user to learn a new solution, necessity of "learning" new technology) or organizational (e.g. related to internal regulations and organizational codes) constraints.

It is often access to infrastructure (both to IT tools and to manufacturing machinery or laboratory equipment) that is key to the selection of project team members. In such cases, there are many independent components of the infrastructure that work together for the project. This cooperation is, to a great extent, the exchange and processing of information, which then forms the basis for decision-making.

Most importantly, however, it turns out that modern network organizations - even those classified by us as highly innovative or "high-tech" – are increasingly relying on publicly available solutions in their operational activities. Advanced systems supporting the organization's operations are the domain of great and traditional organizational structures. Modern innovations are more and more often developed by startups, spin-offs (teams deliberately separated from the traditional structure of large organizations), or by networked consortia carrying out collaborative research and development projects. While members of individual organizations that form a network organization have at their disposal an infrastructure that is part of their parent entity, a variety of tools are used to streamline communication between collaborators. These can be simple, popular and widely available solutions such as email, document sharing, common workspace and data cloud solutions, but they can also be Business Intelligence systems that support data analysis. The use of such solutions requires the decision-maker not only to be highly competent, but also to have access to the possibly fullest and most accurate information, to high-quality data.

RISK AND UNCERTAINTY IN DECISION-MAKING

In line with the classical decision model, it is important to analyze all possible solutions as a result of the decision-making process. On the basis of such an analysis, an optimal decision is made. The implementation of this model requires, however, the possession of adequate and exhaustive information about different available solutions and the consequences they entail. The problem is that decisions made under conditions of certainty, i.e. precisely in light of full information, represent a very small percentage of actual problems in organizations. The vast majority of decisions are made under uncertainty or risk. Hence, there is another behavioral or administrative model that assumes that decisions that are made in incomplete and imperfect information environment are not entirely focused on making an optimal decision but rather on making one that is satisfactory enough. Decision-makers in the behavioral model are, thus, limited in their rationality. These limitations are due to numerous factors, such as: personal experience, knowledge, risk, intuition, or values. In this case, the decision-maker usually chooses the first available option that meets the criteria set out earlier. In the organization, the rationality of dealing in such situations is a choice that will be "locally optimal", which means that it will lead to maximizing the organization's goals with the given constraints, both in terms of information, skills and time. One should remember that the decision-making process should take into account the analysis of all possible conditions, which significantly increases duration of the decision-making process, especially in network organizations. For this reason, IT support tools should be used. As per problems marked by high uncertainty, support may be provided by the already mentioned BI systems or high-performance computer platforms, such as IBM Watson or Oracle. These, however, are the solutions used in research and development projects rather than in small start-ups.

The environment in which organizations operate can be characterized by three dimensions, which are: uncertainty, dynamism and complexity (Figure 1). The degree of uncertainty refers to the fact that some events are easier (deterministic cases) and others more difficult to predict (stochastic, probabilistic cases). Dynamism is how a situation changes in time. Here, too, each event may assume different degree of dynamism – some may be highly variable, while others can be constant. The degree of complexity, meanwhile, is the scale between relatively simple situations (consisting of few elements and unambiguous relations between them) and multi-elemental situations composed of relations that are diverse and difficult to identify

The model proposed by R. Howard (1988) (see Figure 1) allows for an analysis of the environment in which the organization operates and it distinguishes eight ideal types of scenarios (cube vertices), which can also be presented as a decision tree (Figure 2):

Situations near vertices 1, 2, 3 and 4 are deterministic. In such situations (which are certain, sure), the decision-maker deals with risk-free tasks. Here we can predict the situation will be mostly encountered by traditional organizations. Situations near vertices 5, 6, 7 and 8 are stochastic (probabilistic). In these cases, the decision-maker is not certain, sure as to the result(s) of his or her decision. The tasks dealt with here come with a risk, and they are accordingly called risky tasks. An additional difficulty in making risky decisions may be high variability or complexity of the situation or the emergence of both these characteristics. This is most likely to occur in network organizations. The more network elements there are, the more complex is decision-making.

Figure 1. Three-dimensional approach to decision-making

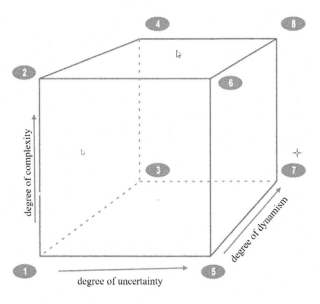

Figure 2. The decision tree

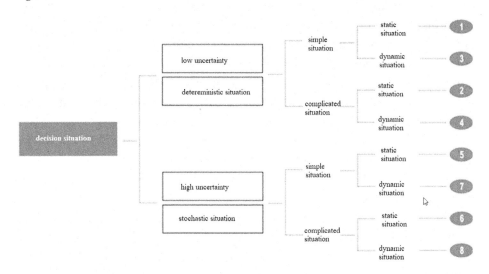

Analysis of the above factors makes it possible to identify four types of decision problems: deterministic, moderately probabilistic, highly probabilistic, and indeterministic.

A problem can be referred to as deterministic when the decision-maker is clear about the goal he or she wants to achieve and knows that applying his or her solution provides certainty in obtaining the expected result. This is, therefore, a problem with one correct solution and one proper way of reaching it. Here we apply optimization.

Moderately probabilistic problems occur when the goal is clear and the finite set of possible ways to achieve it is known, but the result of each of these way is dependent on independent conditions, the probability of occurrence of which can be predicted. Although the decision-maker in this case does not directly control many of the conditions under which he or she is forced to operate, he or she is nevertheless able to predict them, and thus able to determine the probability of achieving the desired result. In this kind of problem, the decision-maker takes the risk, but can still determine its extent and undertake those actions where it will be the least severe. Different types of risk estimation methods and methods of system analysis are used for this kind of problems.

Then there are highly probabilistic problems. In this case, the-decision maker is clear about the goal, knows a limited number of ways to achieve it, but cannot predict what factors and probabilities can occur in the implementation of the decision and determine the success or failure of his or her actions. In such instances, the decision-maker must take significant risks. Aside from the methods mentioned in previous cases, simulation methods are additionally used here. The purpose of simulation is to approximate the reproduction of phenomena or behaviors of a given object. The idea of simulation is a specific "rehearsal" of the problem and its verification, using known solutions of its effects. Simulations are often conducted using information systems supported by mathematical models or algorithms. They are co-developed with the support of professionals, people with extensive experience in the field, access to which would normally be very much hampered. It should be noted, however, that there is a group of decisions that rely in particular not only on the use of IT solutions and the development of computer simulations (e.g. in pilot training), but often on physical model simulations as well. As per working on physical models, it is possible to observe the object and its direct reaction, e.g. to changing atmospheric conditions.

The last group concerns indeterministic problems. Here, the decision-maker has difficulty in determining (i.e. accurately dimensioning) the desired goal. Neither can he or she demonstrate the actions or the probabilities of the factors that may arise by changing the environment in which he or she is to operate. In extreme cases, one can speak of acting under extreme conditions. Does this mean that the subject reacts in an ill-conceived and completely accidental way? Not necessarily. His or her knowledge and skills may, however, fare worse than in the previous cases, but the methods of mathematical analysis, system analysis - including simulation - help understand the problematic situation, determine overall states of the desired changes and predict factors that may arise in the future. All these approaches form part of a broad range of systems that support decision-making.

IT SOLUTIONS SUPPORTING DECISION-MAKING PROCESSES

The development of information technology (IT), as well as the derivative of this phenomenon in the form of changes in management, triggered the tendency to integrate IT systems functioning in organizations. The result is the MRP, ERP, or highly relevant and well-developed business analytics solutions. Selected technical solutions, particularly important from the standpoint of the functioning of the network organization, are discussed in Chapter 3 of this monograph. These systems work both within individual organizations and network organizations. The development of the network has resulted in the opening of organizations and their information resources to the idea of cooperation. In this way, the mechanism that used to take place in individual organizations was repeated. Organizations take advantage of the opportunities offered by the global network to redefine their way of doing business. Engaging in networking and collaborative actions is a process similar to that of integrating information systems at companies. Whereas the first wave of systems integration involved merging and systemizing solutions and data sets within a single organization, today integration puts more emphasis on collaborative networks, many of which are independent organizations.

The purpose of the information subsystem is to collect, store, process and transmit information. This information is then passed on to the decision subsystem where it is used to create decision variations. The final decision in this case does not belong to the system but to the decision-maker. The system is merely a knowledge base that effectively processes the search for solutions to a specific problem. Once the decision is implemented, the stage of solution evaluation takes place, strongly supported by the information system that gathers information about the expected effects of the implementation of the decision. Information on the consequences of the decisions may be the basis for further action.

Information systems dedicated to supporting particularly difficult decision situations, i.e. those for which there are no clear procedures to achieve an optimal solution, are referred to as the DSS (Decision Support System). Their task is to prepare acceptable variations of the decision-making solution and assist the decision-maker in selecting the optimal one from among them.

According to the classification provided by E. Turban (Turban et al., 2015), decision support systems are divided into five key categories, which are:

1. Data-driven DSSs, which rely on data and processes to transform data into information;
2. Model-driven DSSs, where the main focus is on simulation and optimization models;
3. Knowledge-driven DSSs, characterized by the use of knowledge technology in solving specific problems in the decision-making process;
4. Document-driven DSSs, which enable the user to acquire and process data from various, often unstructured documents or web pages;
5. Communications-driven and group DSSs, where technology that promotes communication and cooperation between system users is particularly important.

A different is shared by S. French (Rijos Insua & French, 2010, 67-70), who refers in his proposed classification to the managerial level and context in

Figure 3. The architecture of decision support systems

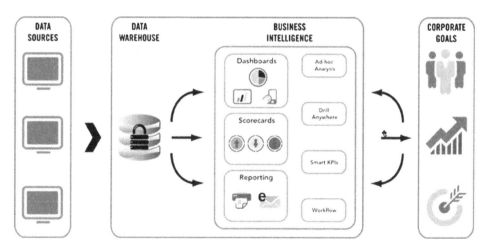

the decision-making, which comes down to strategic, tactical and operational decisions. In this case, decisions made at strategic level are those that set the direction and the framework for decisions made at "lower" levels in the organizational structure. This classical division is, however, supplemented by an additional perspective that was previously highlighted in D. Snowden's work, known as the Cynefin framework (Snowden & Boone, 2007). This framework takes into account four decision contexts, or "domains". These are: simple, complicated, complex and chaotic (see Figure 4). They are meant to give decision-makers a sense of place. Domains on the right side of the model can be both simple (sometimes even obvious) and complex, but they are always characterized by order. Their causes and effects are either known or can be discovered along they way. Much more complex are domains on the left since they are characterized by disorder. In these contexts, causes and effects can be derived from reasoning and analysis, but it takes time (in many cases such inference may be made in retrospect). In extremely difficult cases, resolution of problems within this domain group may end up in failure.

The context grid for decision-making processes, as offered by Cynefin, is supplemented by S. French by the use of decision support systems, noting that each domain requires a different kind of support. Depending on the tools and technology applied, S. French suggests four levels of support for specific contexts of decision problems (Figure 5).

Figure 4. Relationship between the perspectives offered by the strategy pyramid and Cynefin
Source: (Rijos Insua & French, 2010, 67)

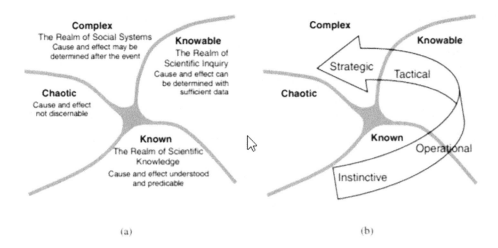

(a) (b)

Figure 5. Categorisation of variety of DSS according to Cynefin space and level support
Source: (Rijos Insua & French, 2010, 69)

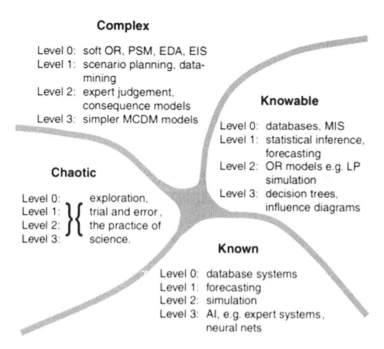

Key: AI artifictal intelligence; EDA – exploratory data analysis; EIS – executive information system; LP – linear programming; MCDM – multi-criteria decision making; MIS – management information system OR – operational research; PSM – problem structuring methods

Modern decision-making systems tend to focus on functionalities such as:

1. **Report Generation Speed:** Short response time on user request;
2. Ability to filter and sort data;
3. Ability to draft multidimensional reports, not only tables but also visually appealing ones (graphical reports);
4. Automation of the reporting process as the ability of the system to perform repetitive reports on its own;
5. Intuitive user interface;
6. Collaboration with other applications used in the organization, such as office suites (word processors, spreadsheets and databases).

Decision problems that are resolved through decision support systems are related to the collection, organization and sharing of management information and organizational knowledge. These systems provide significant support for the quality of stored and shared information, and also provide protection for access to information and knowledge as they allow for the control and matching of user permissions regarding those systems. The ability to access knowledge bases and models has a positive impact on the speed of information that ultimately reaches the decision-maker. It is therefore possible to make decisions not only fast, but also effectively.

DECISION-MAKING IN NETWORK ORGANIZATION ENVIRONMENT

The process of decision-making is most often represented as a series of interrelated phases or activities. Decision-making in which all stages are defined is an algorithmic process and its implementation is an algorithm, i.e. a set of rules of action determining the resolution of a given problem in a finite number of steps. This situation can arise in traditional organizations. On the other hand, in network organizations marked by significant autonomy, heuristic methods (creative-thinking methods) are opted for instead. Heuristic methods need to be able to detect facts and relations between them. Most discoveries, inventions and unconventional methods of proceeding have been achieved through heuristic methods of problem-solving. According to Mary Jo Hatch (2012), a typical rational model of decision-making with control elements through feedback consists of five steps:

1. Identify the problem,
2. Define and evaluate the possible options,
3. Choose one of the options,
4. Implement the chosen option,
5. Observe the results.

The decision-making process can be considered as a decision-making chain. When making the first decision, one should consistently make the following ones. The results obtained after the first decision affects subsequent decisions. Such models of chain decision-making are referred to as dynamic. Despite the diversity of problems and multi-faceted decision situations in organizations, a general algorithm (model) for decision-making situations was developed. Making a decision is the key, but it is not the most important factor given that the ultimate success or failure of the organization is the final result - the degree of achievement of the opportunities that arise in a given situation or reduction of threats, bridging the gap between what is and what should be. The concept of the organizational decision-making process can be understood as typical steps (stages) of actions usually taken by decision-makers. In network organization, we can speak of a collection of them. What is characteristic of the decision-maker is that the goal is not only to make a decision, but to resolve the problem in the process of doing so. The problem-solving process in an organization can be described as a cyclical sequence of related actions taken by decision-makers. The model shown in Figure 6 is postulative (normative).

In reality, it is difficult to separate the stages of decision-making and decision-implementing. In practice, as indicated earlier in this chapter, both the return to the earlier stages and the omittance of some steps are observed. The decision model used is relatively rare. It can be a model for action when the problem faced by the decision-maker is simple, the conditions are unambiguous, the solutions are relatively inadequate, the criteria for evaluating them are not questionable, and the decision-maker has the necessary resources to analyze and verify all alternatives and predict consequences of his or her individual choices.

Figure 6. A model of decision-making in the organization

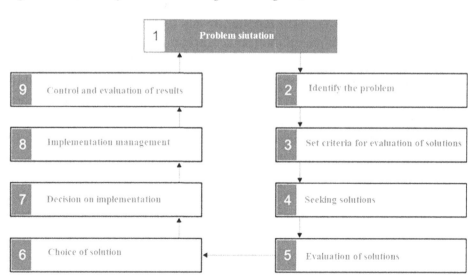

In the case of relatively simple decision problems, i.e. when the problem is well-defined, there are only a few solutions and criteria, and the decision-making procedure is therefore relatively simple. It is enough to evaluate each one (analyze it) in terms of all criteria (taking into account the weight of the criteria) and select the one that with the "highest overall score". The criterion of feasibility seems to be of paramount importance for practitioners (sometimes it assumes the form of the question "Is it realistic in our organization?"). An example of such solution is provided by James A. F. Stoner (1997). The border between the analysis and evaluation of solutions and the choice of one lies in organizational practice, which is sometimes difficult to grasp. Choosing a solution may emerge in the evaluation process. However, when there is no clear "favorite" in the solution evaluation stage, one of the four procedures (ways) of choosing the final solution is used based on the already analyzed and evaluated alternatives: comparing the main parameters of some of the most promising solutions, voting, negotiating and making an intuitive choice. The choice of one of these procedures seems to largely depend on the extent to which decision-makers agree as to the goals they want to achieve, as well as on alternatives and their consequences (effects). According to J. D. Thompson (Hatch, 2012), decision-makers in organizations make decisions by calculation (mathematical models including optimization and simulation), trial and error, or negotiation. In modern organizations, difference of opinions, views or interests is common, while turbulent environments make it increasingly difficult to anticipate the consequences of decisions. In case of disagreement as to the purpose (conflict of objectives), but retaining a consistent position on the methods of action and their possible consequences, the conflict can be resolved by negotiating and working out a compromise. Under the framework of negotiation and compromise model, agreements and common values are reached.

When agreement cannot be reached on how to solve the problem and its consequences, but when there is agreement about the general objectives, then the problem of choosing the decision can be resolved by testing the variations. The test queue for individual solutions may be derived from the analysis of the ratings assigned to solutions at the earlier stage of their evaluation, or if it is difficult to clearly rank, the team of decision-makers can vote on it. The tested solution can be verified by the technique of organizational area separation where the experiment is conducted and the outcome of the solution is observed.

THE PATH FROM THE DECISION-MAKER TO THE NEGOTIATOR

Decisions made within an organization can be either individual or collective. The advantage of network organizations (as described in Chapter 4 of this monograph) is that they abandon the rigid, traditional organizational structure. The traditional structure was clearly related to the existence of a manager responsible for making decisions. With the development of modern and updated organizational structures, the process of gradual expansion of non-managerial employees became apparent. This process was gradual but consistent, and more and more employees had the right to make decisions regarding his or her current activities. Such transfer of managerial functions to employees finds its culmination in a network-based structure whose shape supports the organization's current functions in a fully decentralized manner. It must be remembered, however, that even in the network structure, there is a leader, or a manager who has the final say and who assumes responsibility for the consequences of the decisions that are made. The literature points to a number of advantages and limitations associated with individual or group decision-making. Decisions made by one person can be distinguished by:

1. Greater sense of responsibility of the decision-maker (in the case of individual decisions, responsibility for possible consequences is not blurred, which is a significant problem in the case of collectively-made decisions);
2. Rapidness - decision-making by a group of people involves the need to devote sufficient time to discussions and debates;
3. Lower unit costs for the decision-making process;
4. Less risk of compromise that does not satisfy any of the parties. These are preferred in difficult cases where it is necessary to work under particularly complex circumstances. For this reason, in critical cases with a strong pressure of time (safety, saving human life), decisions are made by one person.

One-man decision-making also involves numerous limitations, which tend to be as follows:

1. Substantially worse, due to the fact that the knowledge of certain professionals will always outweigh the knowledge of one person;

2.   Consequent risk of higher costs associated with making wrong or bad decisions;
3.   The decision-maker may have no group of consultants or rational critics;
4.   Decisions made on a case-by-case basis must be followed by a team of employees (subordinates). In this case, people who are not involved in decision-making often do not demonstrate commitment to their implementation.

Making decisions in network organizations does not necessarily mean collective decision-making. Regardless of whether the organization functions in a traditional or networked structure, there is usually a person who assumes the role of a decision-maker. This is usually a manager, a team leader, or a person who will simply assume responsibility for the consequences of the decisions. In studies conducted in network organizations (research sample described in Chapter 1 of this monograph), key decisions – both for the project as a whole as well as for individual assignments – were made by one authorized person, who was most often the task manager or the project team leader. This was the case for the most important decisions In the case of strategic decisions, due to the fact those decisions were undersigned by one person. However, the decision-making process itself has changed along with the evolution of the way in which individuals are organized, their organizational structure. In the traditional structure, according to Fayol's classical notion, where the unity of command and leadership was of great value, decisions were made on a one-man, authoritarian basis.

Changes in organizational structures (as described in more detail in Chapter 4 of this monograph) - which have become evident through increasingly progressive decentralization and flexibility - emerged alongside the development of successive management approaches as well as changes in decision-making processes. The decision was increasingly made in a collective way and the role of the manager was becoming more and more integrative than any other thing. Which is to say, managers relied in decision-making process increasingly on the opinions and experiences of their subordinates - specialists and the knowledge gained directly from the people involved in decision-making and the available information systems. This is a decision-making period supported by simulations and the use of any kind of support in the form of ICT solutions, models or algorithms. Performing subsequent simulations is the starting point for optimal decision-making, which is more often a result of the cooperation of the involved team.

In today's world marked by an increased presence of network organizations, it becomes each time more difficult to speak of classical organizational relationship. Those who participate in decision-making processes are very often not organizationally organized or hierarchically linked. This dependence has several advantages, since at the end of the day - even in difficult decision environment with various controversies, different views and experiences – there was a need for a final decision to be reached. This necessity was reinforced by the functional dependence of the members of the organization. This element is not particularly crucial, or in extreme cases is non-existent in network-based organizations. Decision-making relying on an organizationally independent group of people representing different areas of specialization and different perceptions may prove much more challenging than in the case of decision-making as it used to be known. In organizations with a network structure, negotiating skills may be a particular element that will determine the success or failure of the decision-making process as a whole. Decision-making in network organizations increasingly resembles negotiation, and in the context of decision-making by peer-to-peer organizations based on a strongly democratic structure, skills and willingness to negotiate are the cornerstone of successful decision-making.

DECISION-MAKING TOOLS IN A NETWORK ORGANIZATION ENVIRONMENT

The evolution described earlier in the approach to decision-making processes is also evident in the results of the research conducted for the purposes of this monograph. As part of the research, the usefulness of the individual IT tools (as described in Chapter 3) in the decision-making processes occurring in the context of the implementation of project activities was identified. Presented here are the conclusions of the surveys conducted on the research sample (also described in Chapter 3 of this monograph). The respondents however, did not answer questions concerning knowledge or use of IT solutions, but rather their usefulness in decision-making during project tasks. From among the respondents who participated in the project implementation as both project managers and a single-task managers, only 4.9% of admitted to not have participated in the decision-making process. That group was made up of people supporting the project, rather than directly managing it, and they were involved in administration, accounting or logistics. In this respect, their participation in the decision-making process was limited because of their function. The other respondents admitted that to have participated in the decision-making process by assuming different roles such as the decision-maker, the specialist consultant or the analyst. The decisions in the analyzed projects were made collectively. The dominant model of decision-making in the analyzed organizations was an analytical model, in which the need for an optimal solution prevailed (Figure 7). The survey questionnaire was used in the classification proposed by Simon and March (1964, 35-38), where two classical ways of approaching conflict are indicated. The cause of conflict in the organization may also be the decision-making process in a network organization where the parties represent distinct perspectives and sometimes other areas of interest. The first two methods rely on analytical procedures, namely persuasions and the search for an optimal solution. In this case, the parties are guided by their respective decision-making know-how, their own experience, and their logic. In the case of decisions made by means of negotiations or politics, analytical elements are not a priority factor. While the first group of solutions assumes a community of goals and interests, and the difficulties involved in decision-making are due to the need to find the optimal solution, the second group puts the main emphasis on the position of

Figure 7. Decision-making in research and development projects

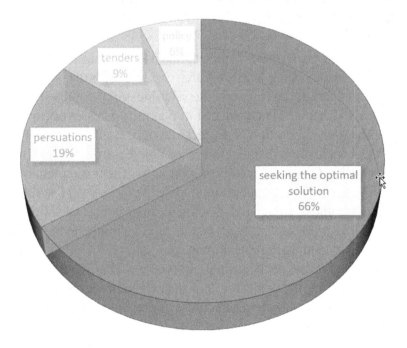

power and authority, and therefore incompatibility of interests between both may be evident . In the case of the analyzed organizations, the second case, i.e. the emerging conflict of interest, emerges, albeit in a much smaller number of responses, as only 15% of the respondents are concerned. In the case of the remaining 85%, decision-making is based on analytical considerations and the search for the best possible solution.

Among the IT tools, widely discussed earlier in Chapter 3, which are particularly relevant for decision-making in network organizations, there were those that enabled the sharing of knowledge. Document-sharing tools (DMSs) that allow communication between distant teams or individuals (video-conferencing) and systems collecting data and knowledge (knowledge basses) were indicated as particularly important. The survey results are shown in Figure 8.

In the aggregate results of the questionnaire, it is clear that a practical solution often applied by teams participating in R&D projects are industry IT systems and other systems for data analysis. The specificity of networking, however, forces organizations to continually engage in dialogue, which is why there is a clear primacy of tools to communicate and support this area of the organization. This group of tools includes those that belong to the "workspace" area, such as Asana or Slack, which provide a common space for exchanging documents (in various formats and shapes) for the members

Figure 8. Utility of IT tools in the decision-making process in network organizations

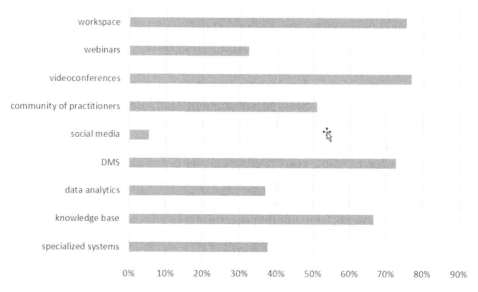

of the collaborating group, storing them, running discussion panels in groups or closed discussion groups and separate design teams. These are the platforms that support the way the project is implemented. Although there is no limitation for the user and the way his or her project is pursued, according to survey results, these solutions are particularly active as support tools for network projects.

Successful communication should support the process of finding the best solution and making the right decision. However, a solution based solely on analytical methods is not always possible. Out of 82% of the respondents, decision-making problems within the framework of cooperation were not resolved unequivocally, using mathematical models or simulations. These problems are characterized by a very high degree of complexity and are related to the inability to find a solution in a way that is based only on data analysis and simulation. In this case, it is necessary to make decisions based on knowledge and experience (know-how). Due to the diversity of experience and the high level of diversity of the persons involved, they participate in the decision-making process by negotiating, or working out, a final decision. Similar results, indicating the importance of negotiations for the functioning of network organizations, were obtained in the group of participants of international projects pursued by university students. The study was a participatory observation, and the networking of the organization was based on the need for cooperation between students representing different fields of knowledge and different geographic, linguistic and cultural academic centers. In the case of student groups, it proves impossible to identify one leading decision-making center. While in the case of the surveyed organizations, one could pinpoint a "leading" unit, or one that has larger (more crucial) resources in the project, in the student group the division of tasks and qualifications is completely uniform. Although individual project team members have a different conceptual and other knowledge, their importance and role in the organization is the same. Solving subsequent tasks depends, then, on the necessity of dialogue and working out a satisfactory solution, precisely through negotiation which leads to a decision that breaks down different approaches and accounts for existing constraints. This is not a classical compromise since, as evident in the work of student groups, scenarios arise when groups representing one of the disciplines must introduce radical changes under the influence of arguments from another discipline. In any case, however, this is a way of gradually solving the problem. As with the surveyed group, tools that are indicated as the foundation for communication and decision support are IT tools - mostly "workspace" tools, instant messaging and social networking.

CONCLUSION

Along with the transformation regarding the way of achieving organizational goals, the decision-making process itself is being modified. Classical approaches derived from management sciences point to the essence and unity of the manager as a superior planner, organizer and decision-maker. Over time, this approach has gradually changed. In process-oriented organizations, the decision was also made by the decision-maker, but alternative solutions relied on simulations and optimization. These days, more and more project organizations operate under uncertainty. Attempting to simulate or optimize processes proves, in the absence of knowledge of the process and the shape it assumes, impossible. This uncertainty requires organizations to make a complete shift of their perspective, consistent with non-optimal decisions and those that will result in a consensus. The decisions made by today's organizations are ones that stem from negotiations. Network organizations made up of diverse collaborators must include in their actions the need to meet the requirements of each party. The lack of such organizational *empathy* can result in the breaking apart of a network organization and the abandonment of cooperation. Hence the necessity of negotiation. Negotiations are not just about working out common solution that is often not the optimal one, but merely "good enough." It is also an opportunity to declare one's own standpoint, strengths and limitations. Negotiations in this case may provide a starting point for further project collaboration, thereby allowing for a better understanding of the collaborator and for building trust among the various members of the network organization. Negotiating in decision-making situations for network-based organizations is a discussion conducted among peer partners. It is important to note, however, that the equivalence of these partners may not be obvious for specific tasks. Depending on the type of activity being undertaken, each member of the network organization enjoys either more or less qualification and competence. For this reason the bidding power of each party may vary across the decision-making process. Therefore, in the case of network organizations – which, by definition, foresee the equality of partners - there is, in practice, a paradox of inequality. Decisions are no longer just a compromise, but rather an area of discussion and a place to defend one's "bargaining power". Consequently, negotiations are not only to make good decisions, but also to pursue the goals and interests of all members of the network organization.

REFERENCES

Adair, J. (2011). *John Adair's 100 greatest ideas for smart decision making*. Chichester, UK: Capstone Publishing Ltd.

Beckers, T., & Craske, M. G. (2017). Avoidance and Decision Making: Implications for the Understanding and Treatment of Anxiety. *Behaviour Research and Therapy*, *96*, 1–106. doi:10.1016/j.brat.2017.05.009 PMID:28545650

Blake, C. (2010). *The art of decision. How to manage in an uncertain word*. Pearson Education Inc.

Bojar, W., Rostek, K., & Knopik, L. (2013). *Systemy wspomagania decyzji*. Warszawa: PWE.

Davenport, T. H. (2011). *Making Smart Decisions*. Harvard Business School Publishing Corporation.

Davis, I. (2005, May 26). The biggest contract. *The Economist*.

Forbes. (2017). W UE będą specjalne przepisy dotyczące robotów i sztucznej inteligencji. *Forbes*. Retrieved from: https://www.forbes.pl/technologie/przepisy-dotyczace-robotow-i-sztucznej-inteligencji/rwth4xw

Gaynor, G. H. (2015). *Decisions. The Engineering and Management Perspective*. The Institute of Electrical and Electronics Engineers Inc.

Guardian. (2014). Eugene the Turing test-beating 'human computer' – in 'his' own word. *The Guardian*. Retrieved from: https://www.theguardian.com/technology/2014/jun/09/eugene-person-human-computer-robot-chat-turing-test

Gunther, R. E. (2008). *The truth about making smart decisions*. Pearson Education Inc.

Gunther, R. E. (2011). *The truth about your emotions when making decisions*. Pearson Education Inc.

Howard, R. A. (1988). Decision Analysis: Practice and Promise. *Management Science*, *34*(6), 679–695. doi:10.1287/mnsc.34.6.679

Hujer, T. (2011). *Design and Development of a Compound DSS for Laboratory Research, Efficient Decision Support Systems - Practice and Challenges From Current to Future.* InTech. Available from: https://www.intechopen.com/books/efficient-decision-support-systems-practice-and-challenges-from-current-to-future/design-and-development-of-a-compound-dss-for-laboratory-research

Kisielnicki, J., & Zach, R. (2012). Decyzyjne systemy zarządzania (DSZ) – pojęcia, modele, procedury. In J. Kisielnicki & J. Turyna (Eds.), *Decyzyjne systemy zarządzania* (pp. 13–54). Warszawa: Difin.

Kubanek, J. (2017). Optimal decision making and matching are tied through diminishing returns. *Proceedings of the National Academy of Sciences of the United States of America, 114*(32), 8499–8504. doi:10.1073/pnas.1703440114 PMID:28739920

Laudon, K. C., & Laudon, J. P. (2018). *Management Information Systems: Managing the Digital Firm.* Pearson.

Luoma, J. (2016). Model-based organizational decision making: A behavioral lens. *European Journal of Operational Research, 249*(3), 816–826. doi:10.1016/j.ejor.2015.08.039

Ríos Insua, D., & French, S. (Eds.). (2010). e-Democracy, Advances in Group Decision and Negotiation. Springer Science+Business Media B.V.

Sauter, V. L. (2010). *Decision Support Systems for Business Intelligence.* Wiley.

Simon, H. A. (1960). *The New Science of Management Decision.* New York: Harper Row. doi:10.1037/13978-000

Simon, H. A. (1995). A behavioral Model of Rational Choice. *The Quarterly Journal of Economics, 69.*

Simon, H. A., & March, J. G. (1964). *Teoria organizacji.* Warszawa: PWN.

Snowden, D., & Boone, M. (2007). A leader's framework for decision making. *Harvard Business Review, 85*(11), 68–76. PMID:18159787

Stanek, P. (2013). Theoretical Aspects of Collective Decision Making - Survey of the Economic Literature, Comparative. *Economic Research Journal, 16*(1), 103–120.

Turban, E., & Aronson, J. E. (2015). *Decision Support and Business Intelligence Systems*. Pearson Education.

UNIDO. (2017). *What is CSR?* United Nations Industrial Development Organization. Retrieved from http://www.unido.org/crs/o72054.html

Walczak, W. (2012). Czynniki i uwarunkowania wpływające na decyzję w zarządzaniu organizacją. *eMentor,* 3(45). Retrieved from: http://www.e-mentor.edu.pl/artykul/index/numer/45/id/933

Chapter 6
Network Organization as a Result of the Strategy of Forming Global Organizations:
Organizational Structure Transformations

ABSTRACT

In this chapter, the authors assume that if an organization implements a development strategy supported by the information communication technology (ICT) progress, its organizational structures are transformed. The implementation of such a strategy leads to the transformation of local organizations into global ones. Organizational structures, forming part of the infrastructure for the implemented strategy, change. As a consequence, the transition from hierarchical structures to network structures takes place. The transformation results in the rise of global organizations.

INTRODUCTION

Based on research in American corporations, D. Chandler (1962) depicted interdependences between the organizational strategy and structure. The analyses of development of firms such as Du Pont, General Motors, Standard Oil Company, and others indicated that the organizational structure should be aligned with the implemented strategy (structure follows strategy). D.

DOI: 10.4018/978-1-5225-5930-6.ch006

Chandler's research on the evolution of organizational strategies and structures of American corporations revealed close relationships between organizational strategies and structures (Kisielnicki & Nowicki, 1974). At the initial stage, organizations focus their activities on one product. Over the course of time and their development, their structures evolve. The changes relate to new tasks resulting from their growth. A change of strategy very often involves diversification and emergence of new departmental forms or mergers with other organizations. At the beginning of life of organizations, management decisions were centralized in the hands of one manager or a small management board. The reason was that the degree of complexity and formalization of the structure was low. The development of organizations and their increased role changed their strategies. This, in turn, required more complex management structures and decentralized hierarchies.

NETWORK STRUCTURE AS A METHOD FOR BUILDING A GLOBAL ORGANIZATION

The research initiated by D. Chandler was continued by B.R. Scott (1973), R.P Rumelt (1974), P. Drucker (1977), R.E. Miles and C.C. Snow (1978), H. Mintzberg (1997), J.L. Boschken (1990) and many others. In this respect, B.R. Scott's (1973) observations were interesting. He presented the following path of organizational structure evolution. The initial model was simple with slight formalization – the so-called *one man show*. The next stage was a bureaucratic model that was transformed all the way into a decentralized model. The analysis of related literature and own research on changes in organizational structures may help identify the stages of that evolution. Network organizations are an effect of such evolution of organizational structures.

Stage I: Local Organization With Traditional Management Infrastructure

This stage covers the period before the widespread application of IT (Information Technology), notably the Internet. Organizations were originally formed as a result of the vision of the owner, entrepreneur. Organizational structures were not complex and the existing communication systems allowed direct supervision of each employee. In such circumstances, the structure is simple and not extensive. In this period, the organization develops gradually. Consequently, some of these simple structures must be replaced by more complex ones. This is, among others, because organizational development creates the need for the owner-entrepreneur to be replaced by a team of managers. Specialist functional divisions are established. Although increasingly complex, organizational structures are present in an organization located in one place. The lack of adequate ICT (*Information Communication Technology*) infrastructure determines organizational development and changes in the organizational management process. The research by the team of L.G. Love, R.L. Priem, G.T. Lumpkin (2002) found that organizations have common problems ensuing from ineffective strategy implementation. While implementing the strategy, managerial staff encounters problems with the applied communication model. Executive employees do not understand the tasks they should do as it is difficult to implement changes with a traditional management infrastructure in place. This situation requires such organizational structures that allow for a decentralized management system. In order to implement new strategies, new models of organizational systems and thus a new communication model had to be introduced. Communication models and the operation of filters deforming the information being sent are discussed in greater detail in Chapter 4 on communication and information transmission within network organizations.

Stage II: Network Organization

For a network organization to work, a developed Information Communication Technology (ICT) infrastructure, including computer networks, must be in place. Organizational strategy evolves towards activities in multiple sectors, a variety of products and services offered, and new markets to enter. Such a strategy cannot be followed under the current model of linear and functional structures. As a result, functional models are transformed towards network models of organizational structures. In emerging network organizations, separate business units exist such as cost centers and profit centers. These units are the nodes of network organizations. They have great autonomy and are responsible for the results of their activities (Morgan, 1997). The emergence of a network organization can be considered in the following areas:

1. Management structures within organizations,
2. Processes within organizations,
3. Relations with the environment.

All these areas determine the management structure of a network organization.

Stage III: Global Network Organization

Global network organizations typically have advanced IT infrastructure for management and use such modern solutions as cloud computing and Big Data technology. The development of an organization and its management system, as demonstrated by J.R. Galbraith, D.A. Nathanson, R.K. Kazanjian (1986) to name a few, is associated with its internationalization. Today's global organizations are multinational and flexible (*agile*) in order to perform their tasks. Such diversified organizations should use extended network structures. A broad range and scope of management should be supported by modern ICT (*Information Communication Technology*). At this stage of development, organizations adopt a flexible network model. Network structures expand or shrink as needs arise. Highly popular ways of organizing collaboration among companies include joint ventures, just in time, and alliances.

The presented division of management structures is based on the technological approach to the evolution of structures. The opponents of this division may allege that the primary source of change, as claimed by D. Chandler, is a change of strategy. However, new organizational structures could not emerge without changes in the management infrastructure (Kisielnicki, 2016). As written by A.M. Shah (2005), although organizations understand the need for a strategy, they often face many obstacles in implementing it. With that in mind, A.M. Shah attempts to indicate barriers to strategy implementation. His research focuses on identifying factors that help promote effective strategy implementation. Based on the data collected from 104 managers of large enterprises, he demonstrated the obstacles most often encountered by companies when implementing the strategy. As argued by A.M. Shah, they may be removed provided that effective IT systems are used.

The evolution of organizational structures can be exemplified by the SYNAT project in Poland (Kisielnicki & Gałązka, 2012). That project was designed to create a universal, open communication platform for knowledge resources for science, education and an open knowledge society. It was financed from the state budget by the National Center for Research and Development. The program was carried out by seventeen scientific institutions. As a result of its implementation, separate organizations formed a network organization. The emergence of the network organization is shown in Figure 1.

Figure 1. Two-level uprising Polish national organization for the development of scientific and technical information

1. Independent centers for the development of scientific and technical information

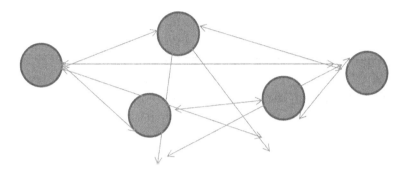

2. National network organization for the development of scientific and technical information

Figure 1 illustrates the process of linking separate academic centers for the development of scientific and technical information into a single coherent network organization under SYNAT. In fact, the project involved 17 academic centers in its first stage. Its future implementation goes towards network links with most academic centers in Poland and connections with global networks for the development of scientific and technical information in other countries in the first stage of that EU project.

An interesting example of the transformation of local organizations into a global network is offered by M. Townsend, L. Coen and K. Watson (2017). They present the transformation of Bayer CropScience (BCS), an international agrochemical subsidiary of the Bayer Group based in Monheim am Rhein (North Rhine-Westphalia). Its turnover is over EUR 12 billion a year. The company was set up during the restructuring of Bayer AG in October 2002. Bayer Crop Protection (founded in 1924) then merged with Aventis CropScience (ACS) acquired in 2001. Bayer CropScience has over 90 subsidiaries worldwide and its operations cover three main areas: Crop Protection (CP), Environmental Science (ES), Bio Science (BS). Local organizations that are part of the BCS did not play a major role in the international market while operating in their headquarters, regions and individual countries. Organizations worked independently, without resorting to profound and extensive experience throughout the trading system for crop protection products[1]. According to Bayer's managerial staff, their characteristics included: slow information flow, late notification of developments, and inconsistent provision of information from regions where public policy priorities were part of the operations. Without clearly aligned priorities, goals, strategies and roles, BCS managers were unable to organize fast and effective action to meet new challenges and be competitive. Realizing that the situation was giving rise to unpredictable and increasingly difficult divergences and risks for the corporation, BCS leaders sought to build a global organization that would consistently and effectively face the challenges of a new external policy. In 2013, they hired Lisa Coen (co-author of the cited article) as a new boss. Her task was to transform the organization with a traditional management structure into a truly global organization. The goal was to continue what had already been good but also to build an interlinked and coordinated global organization. The transformation involved changing the old structure – the existing loose confederation of regional and local offices – into a new structure of interconnected organizations. In this way, a new network structure of a global organization was established. According to M. Townsend, L. Coen and K. Watson (2017), the current network structure allows employees to maintain good local contacts, while encouraging people to share both non-confidential information and information about ideas with people from different regions (such information is often referred to as confidential). Managers were responsible for: communicating the vision, encouraging the sharing of information and knowledge, matching people with relevant information, and mediating disputes.

As written by P. Plastrik and M. Taylor (2006), organizations adopt network structures when they need flexibility, responsiveness and resistance. As opposed to rigid hierarchical structures, networks can rapidly adapt to changing conditions. They can quickly discuss recommended actions, solve new problems and address new risks. As written in the summary of this example, L. Coen, based on current and emerging problems of BCS, designed and implemented an agile organization ready to swiftly predict and respond to problems. The management model that L. Coen created for BCS was a global network organization. It is shown in Figure 2. Another type of a global network organization works in IKEA Retail (Figure 3).

Figure 2. A flowchart of information flow in a global organization
Source: (Townsend et al, 2017)

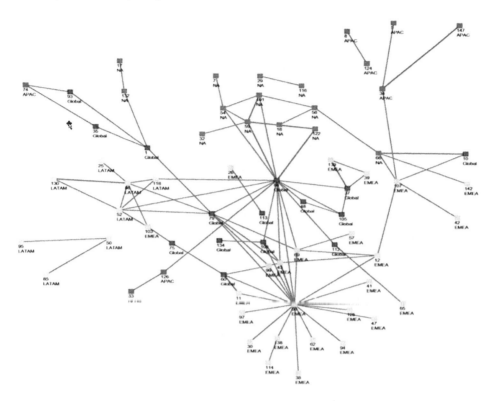

Group Retail comprises retail organizations that are located in specific countries. That is why the figure depicts the division into regions based on the example of Poland and Russia. The region has national managing organizations (national headquarters) in individual countries. They report to Group Retail on the basis of information from each Shopping Center.

Such a structure is more centralized than the BCS structure. This is because a uniform pricing policy must be adopted throughout the company in order for the sales function to be fulfilled. The rule is that individual shopping centers should not compete with each other in terms of product price. It should be noted that different tasks are assigned to Group Industry and Group Product. Their organizational structure is designed for product development and has similar characteristics to the Bayer Group's structure described above.

Figure 3. IKEA Group Retail organization chart

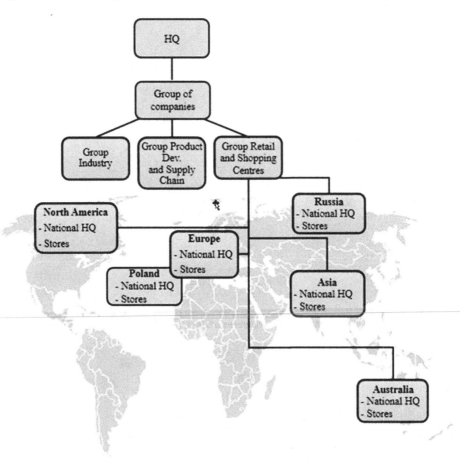

BUSINESS MODELS OF NETWORK ORGANIZATIONS

The concept of business model appeared already in the 1950s in the work by R. Bellman and C. Clark (1957), yet it was not until the 1990s that the theory and discussion about this concept began to develop. Initially, business models were generally applied to analyze commercial organizations. Organizations operated according to the shopkeeper model, whereby a shop was opened where market analyzes indicated that prospective consumers could make a purchase.

The very word "model" is intended to show fundamental assumptions of activity, logic, a certain pattern of behavior, selection of basic resources, scope and nature of internal and external relations, and the main causal relationships among model components. A model is supposed to reflect the part of reality that we are interested in, disregarding less important elements of this reality. It is therefore a construction, a scheme the purpose of which is to indicate, organize and explain a selected section of reality. At the same time, keeping in mind the requirements imposed on us by management sciences, a model must support and enhance control processes. A popular definition of business model proposed by A. Afuah and C.L. Tucci (2011) assumes that a business model is a company-adopted method of increasing and utilizing resources in order to provide customers with products and services that outweigh the competitors' offer and that simultaneously ensure the company's profitability. A similar approach can be seen in other definitions:

1.   A business model describes what a business does and how a business makes money doing those things (T. Malone);
2.   A business model is the method of doing business by which a company can sustain itself (M. Rappa);
3.   A business model describes the rationale of how an organization creates, delivers, and captures value (A. Osterwalder).

To simplify, we can say that a business model is to describe how we should change the management system of an organization so that it benefits the owners or how the system should be financed. It can be assumed that the application of such a business model and calculations led to the transformation of Bayer CropScience into a global network organization.

In practice, different types of network organizations exist. The decision to create and develop them needs to be confirmed by the results of calculations within the business model. Decisions to transform structures are made in view of the turbulence of organizational environment. As shown by Chandler's research, changes in the management system involve improving the development strategy. In turn, this strategy results from actions aimed at building competitive advantages of organizations. The consequences of changes are reflected in recommendations of organizational business models. These models are meant to support decisions about the transformation and formation of network organizations. A business model as a tool for decision-making presents the ways of achieving satisfactory economic results expressed by turnover, costs, and profits within an organization. A business model as a management supporting tool is a static and dynamic model that includes a specific business idea. It is perceived as an instrument that allows verifying whether our business vision will make it possible to obtain satisfactory economic results. It permits an analysis of strategic behavior, taking into account the impact of competitive environment. In practice, there is a myriad of business models. Each organization, depending on its individual needs, generates a model that it needs so that the model takes into account the organization's specifics, internal strengths, and constraints.

Depending on the situation, there are general universal business models or industry-specific models. As a result of the development of the Internet, significant changes can be seen in the design of business models of network organizations. Their application and development allows managerial staff to comprehensively obtain information about new research, new publications, products and services produced worldwide. The Internet has given rise to many diverse models as well as the development of the concept itself. Examples of business models include (Afuah & Tucci, 2001; Rappa, 2003; 2004):

1. **Brokerage Model:** Companies earn commissions on transactions concluded through them,
2. **Advertising Model:** Companies earn fees collected from organizations that advertise their research services and research results,
3. **Merchant Model or Manufacturing Model:** Companies earn from direct sales of project results,
4. **Subscription Model:** Companies earn fees for access to the content published on websites, often highlighting free content for everyone and paid content for subscribers,
5. **Utility Model:** This is a variation of the model presented above. There is no regular subscription fee, and the company charges for the actual use of the service.

In the literature, still newer business models are emerging as market needs arise. Business models whose application seems promising for analyzing the legitimacy of a decision to build a global network organization include:

1. The model developed in 2005 by the team: S. Voelpel, M. Leibold, E. Tekie, G. von Krogh, who believe that a business model is a value proposition to customers and a configuration of a value-delivery network consisting of its own strategic capabilities and other values in that network (e.g. outsourcing, alliances), and the company's ongoing commitment to change and attainment of stakeholder objectives.

2. The model developed in 2007 by the B. de Wit and R. Meyer, who consider that business systems used by organizations are: resources (input), operations (processing), and offered products or services (end product) generating value for customers.

3. The model developed in 2009 by: O. Lisein, F. Pichault, J. Desmecht (2009), who claim that a business model is described by the following three characteristics (axes):

   a. Who are the customers and what type of customers does the organization especially target ?

   b. What are the products/services offered by the organization? What needs do they satisfy?

   c. How does the organization distribute its products and how does it do it better than the competition?

4. The model recommended by D.J. Teece (2010), who considers that a business model articulates the logic of creating and delivering value to customers by business. It specifies the revenue, cost, and benefit architecture of the business that delivers this value. A business model defines how an organization creates and delivers value to customers and how it transforms the reward received into profit.

Synthetic but ambiguous models were proposed by:

1. H. Itami, K. Nishino (2010), who think that a business model consists of two components: a business system and a revenue model.

2. Osterwalder and Y. Pigneur (2002 and 2010), who claim that a business model rationally describes how an organization creates, delivers and captures value.

It should be emphasized, however, that in 2002 those authors provided a more extensive definition of a business model. They wrote that a model represented the conceptual and architectural implementation of a business strategy and the basis for the introduction of business processes. According to them, a business model is an organization's value proposition for one or more customer segments and the architecture of the company and its partner network for creating value and relational capital. The goal is to generate beneficial and sustainable revenue streams.

EXAMPLE OF A BUSINESS MODEL OF
A NETWORK ORGANIZATION

This example concerns the SYNAT project mentioned above. The project aims to create a nationwide network organization with the task of building a platform for hosting and communication of networked resources of knowledge for science, education and an open knowledge society. The approach of A. Osterwalder and Y. Pigneur (2010; 2013) was applied to the SYNAT business model. It is a tool relevant for the class of projects whose products and services are of interest to all partners and stakeholders (see Figure 4), whereas the underlying costs of building and maintaining the platform (here, the SYNAT platform) are covered by public funds.

As reported by A. Osterwalder and Y. Pigneur (2010), the freemium approach was developed by J. Lukin and popularized by F. Wilson. The latter wrote about business models in the online business that combined free basic services and paid extra services. The freemium model involves a large number of users who use the free offer without any obligations. Such a situation exists in a publicly accessible global organization delivering information and knowledge. Such an organization is formed under the SYNAT project. Building a business model is meant to show whether the emerging organization that maintains such a platform is economically viable. Access to the worldwide scientific and technological output is very important for the scientific and economic progress of any country. The organization being created covers a wide range of research tasks, with the overarching goal of establishing a global network organization for Poland's scientific and technical information linked with systems of other countries. In the first stage of globalization, these will be the European Union countries. The tasks of the newly established network organization include:

1. To create an IT platform supporting a wide range of contents, ensuring high scalability and interoperability internationally,
2. To support scientific communication models in open knowledge communities, including a publicity and promotion program addressed to the general public,
3. To create tools that enable the development of new open communication models in science, education and cultural heritage,
4. To provide operational service that ensures system sustainability and to address possible areas of system commercialization.

SYNAT is an attempt to provide knowledge on an open-access basis. In the cited study by A. Osterwalder and Y. Pigneur, an example of a freemium business model is the Red Hat R&D project concerning the development of open-source software.

The conceptual business model of a created network organization is depicted in Figures 3 and 4. The business models shown in these figures make it possible to perform calculations that are relevant for creating a network organization. The quality model in Figure 3 is the starting point for the calculations. It provides information about the data needed for them. ADONIS is a tool supporting business process management based on BPMS (Business Process Management Systems) framework. It was devised at the University of Vienna by the team led by D. Karagiannis (2016). The software used in the business model design was developed at BOC.[2]

The methodological basis for the model was the freemium concept under the approach of A. Osterwalder and Y. Pigneura (2002 and 2010). The quantitative analysis so performed allows the inter-relationship among the categories depicted in the figures to be identified. The calculations made it possible to estimate that the network organization being established will not generate sufficient operating revenues to cover its operating costs in the

Figure 4. The SYNAT business model according to the freemium concept and its adaptation in the A. Osterwalder & Y. Pigneur's approach

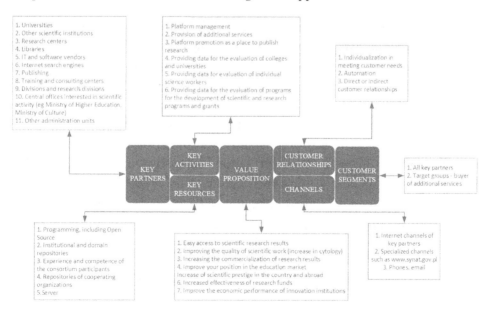

short, and probably even in the long, run. Public funding must be the primary financing source, complemented by off-budget money. When recommending a multi-annual sponsorship program as a means to provide funds for the launch and development of the SYNAT platform, the final and interim targets should be set for each stage of organization formation. These stages may be included in the implementation schedule as, for example, the degree to which repositories are open or the relation of the number of open-access publications to the number of publications resulting from grants financed by public funds, or both targets and their implementation measures in particular years. The calculations performed by means of the business model allowed a road map to be developed with a view to building the SYNAT platform and a network organization operating it. They were made in the question-answer mode. Hence, when decision makers asked the question: What will we get for investing money, designers could quite accurately say what the expected effects would be. At present, such an organization is formed by 17 information centers of various academic institutions. These institutions have double reporting obligations: towards the home institution and the network organization formed by ICM UW (*Interdisciplinary Center for Mathematical and Computer Modeling at the University of Warsaw*).

OPTIMIZATION OF ORGANIZATIONAL STRUCTURES IN NETWORK ORGANIZATIONS

Establishing global organizations requires appropriate organizational structures. When defining criteria for optimizing network organization structures, factors (determinants) that influence them should be identified. The structure of a network organization management system takes the forms of a network of relations among the system elements and the properties of these relations. Today's network organizations are part of an information and knowledge economy. Based on this assumption, the optimization criterion should take into account the information and knowledge transfer as the foundation for evaluating a network structure. A network structure combines things and processes. Organizations constantly look for factors that have a positive impact on their competitive advantage. The use of modern ICT tools such as Business Intelligence (BI) helps support management processes. BI systems make it possible to monitor changes in a competitive environment. New factors constantly emerge, both causing threats and offering opportunities to gain competitive advantage. Autonomous individual organizations that are competitive in the local market are mostly unable to become global market leaders. They should persistently seek sources of competitive advantage both beyond various local organizational structures and abroad.

In analyzing the optimization of network organization structures, we should answer the following questions:

1. Does an organization operating regionally, within the territory of a single country, have such information and knowledge of its competitive advantage that ensures its effective and efficient development?
2. Is a global organization more effective and efficient?

Quantitative answers to these questions are difficult. They are very often impossible with the present state of our knowledge. It is easier to answer the question about the organizational structure that is used in a newly established global organization. We can put forward a thesis about greater effectiveness and efficiency of a network system as compared to a hierarchical structure. As noted by M. Kostera (2014), organizational structure cannot be rigid and should constantly adapt to unfolding changes. In order to operate, a global organization needs space that can only be provided by a network structure. Network structures are the response to the changes in today's global world that cannot be supported by traditional and rigid hierarchical organizational structures.

In the first stage of analysis, the network organization management system should be characterized when searching for optimization criteria. Network nodes, individual organizations, can be quantitatively specified by evaluating the span of control and the number of management levels. The characteristics of an entire network organization are more complex to define. The design of the system of measurements that quantify the whole network was considered by W. Chakon (2012, 2013). He believes that the size of a network is determined by the number of nodes, their density measured as the number of links, the heterogeneity measured as the diversity of nodes and links. The position of a particular company in the network may also be studied. For instance, this is the case of IKEA (Figure 4). Given the need for a centralized pricing policy, a hierarchical organizational structure offers an advantage over a network structure. For IKEA, the peripheral position of the organization (at the ends of the graph of relations) reduces its development opportunities, although it can often protect it against some threats.

Drawing on the research by G. Hamel and C.K. Prahalad (1994) on smart organizations, we can assume that a network organization may be assessed from the following point of view. The optimal structure of a network organization should be organized in such a way that it can support the ongoing learning process throughout its existence. This process involves analyzing the external and internal environment, developing perceptions of the environment, giving the meaning through interpretation, and taking action and correcting competitive behaviors.

C.K. Prahalad and M.S. Krishnan (2008) found that contemporary organizations do not link strategies with building organizational capacities that enable organizations to achieve and maintain continuous change and innovation. Those studies demonstrated that the observations contained in the above cited A.D. Chandler Jr.'s work (1962) are still valid. Network organizations are faced with requirements that should, as written by C.K. Prahalad and M.S. Krishnan (2008), create conditions for access to the global resource network so that unique experiences are created together with customers. To this end, CEOs, executives and managers at all levels of network organizations need to transform their business processes, technical systems, and supply chain management by implementing social and technological infrastructure in order to gain a permanent innovation advantage.

Actions intended to satisfy the requirements imposed on today's network organizations are supported by BI systems. These systems, including their knowledge bases, were discussed in the previous chapter. Here, we would like to note a supportive role in the operation of network organizations that is played by additional knowledge provided by artificial intelligence (AI). In related literature, this issue is addressed by M. Tegmark (2017), a physicist and co-founder of the Future of Life Institute in Cambridge, Massachusetts. The Institute's motto reads: *"Technology is giving life the potential to flourish like never before ... or to self-destruct."* In his book "Life 3.0", M. Tegmark argues that the past development of the world, including organizations, is associated with the evolution of artificial intelligence (AI). Life, says M. Tegmark, may spread throughout our universe and "flourish for billions or trillions of years" because of decisions that we are making now. Artificial intelligence applications are complemented by new computer capacities. The combination of these two trends will allow a new quality of network organizations to be achieved.

M. Tegmark takes into account risks and benefits alike. The disappearing risks encompass both an arms race for autonomous weapons and a radical reduction in employment. The AI community is virtually unanimous in condemning the development of machines that may decide to kill people, yet the issue of replacing human work by devices is provoking a debate. Many predict economic benefits. AI inspires creation of new jobs to replace old ones, as in the case of previous industrial revolutions. However, the following question remains: what kind of desirable economy could we seek when most of what we now call work is done by machines? The actual problem lies in inappropriate goals. M. Tegmark believes that it is unclear how such goals can be pursued by means of purely artificial intelligence that is undefined and may lead to the elimination of mankind. Our task involves the aim of optimizing organizational network structures as a developmental and future-oriented exercise in our monograph

CONCLUSION

Based on our own research and the cited literature, we can specify that the optimal network organization structure should take into account the following requirements:

1. The application of information communication technology with solutions such as big data, cloud computing, smart platforms that should create a management architecture forming the basis of the network organization operation;
2. Management systems allowing for both co-creation of new values with customers and coherent interoperability of all elements of a network organization;
3. Flexible actions to continuously improve the effectiveness and efficiency of a network organization as a whole;
4. A steady flow of knowledge among all stakeholders such as employees, investors, suppliers, customers, which is connected with work in different cultures and time zones in a single global network;
5. Building teams that can quickly deliver high-quality solutions based on knowledge gained from working in network teams.

A network organization must continually propose new innovative solutions and pursue network organization transformation processes to meet the information needs of shareholders, employees, consumers, and interested groups of the information society.

REFERENCES

Anderson, C. (2009). *The Future of Radical Price*. London: The Random House.

Afuah, A., & Tucci, C. L. (2001). *Internet Business Models and Strategies: Text and Cases*. Boston: Mc Graw-Hill Irwin.

Bellman, R., Clark, C., Malcolm, D. G., Craft, C. J., & Ricciardi, F. M. (1957). On the construction of a multi-stage, multi-person business game. *Operations Research*, 5(4), 496–503. doi:10.1287/opre.5.4.469

Boschken, H. L. (1990, March). Strategy and Structure: Reconceiving the Relationship. *Journal of Management*.

Czakon, W. (2012). *Sieci w zarządzaniu strategicznym*. Warszawa: Wolters Kluwer Polska.

Czakon, W. (2013). Splątanie gospodarki. *Być, mieć czy władać, Zeszyty Naukowe Wyższej Szkoły Bankowej w Poznaniu, 49*(4),

Chandler, A. D. Jr. (1962). *Strategy and Structure: Chapters in the History of the American Industrial Enterprise*. Cambridge, MA: MIT Press.

Drucker, P. (1997). *The Best of Peter Drucker on Management*. Boston: Harvard Business School Press.

Galbraith, J. R., Nathanson, D. A., & Kazanjian, R. K. (1986). *Strategy implementation: Structure, systems and process*. St. Paul, MN: West Publishing.

Hamel, G., & Prahalad, C. K. (1994). *Competing for the Future*. Boston: Harvard Business School Press.

Itami, H., & Nishino, K. (2010). Killing Two Birds with One Stone. Profit for Now and Learning for the Future. *Long Range Planning*, 43(2-3), 364–369. doi:10.1016/j.lrp.2009.07.007

Karagiannis, D., Mayr, H. C., & Mylopoulos, J. (Eds.). (2016). *Domain-Specific Conceptual Modeling, Concepts, Methods and Tools*. Springer Verlag Switzerland. doi:10.1007/978-3-319-39417-6

Kisielnicki, J., & Nowicki, R. (1974). *Ewolucja strategii i struktury organizacyjnej korporacji amerykańskich*. Warszawa: Instytut Organizacji Przemysłu Maszynowego.

Kisielnicki, J. (2014). *Zarządzanie i informatyka*. Warszawa: Placet.

Kisielnicki, J., & Gałązka-Sobotka, M. (2013). *Rozwiązania organizacyjne zapewniające trwałość systemu informacji naukowo- technicznej*. Warszawa: Wydawnictwo Uczelni Łazarskiego.

Kostera, M. (2014). *Occupy Management! Inspirations and ideas for self-management and self-organization*. Oxford, UK: Routledge.

Lisein, O., Pichault, F., & Desmecht, J. (2009). *Les business models des sociétés de service actives dans le secteur Open Source. Systémes d'Information et Management*, 14(2).

Love, L. G., Priem, R. L., & Lumpkin, G. T. (2002). Explicitly Articulated Strategy and Firm Performance Under Alternative Levels of Centralization. *Journal of Management, 28*(5), 611–627. doi:10.1177/014920630202800503

Miles, R. E., & Snow, C. C. (1978). *Organizational Strategy, Structure and Process*. New York: McGrawHill.

Morgan, G. (1997). *Images of Organizations*. SAGE Publications.

Osterwalder, A., & Pigneur, Y. (2002). *An e-Business Model Ontology for Modeling e-Business*. 15th Blend Electronic Commerce Conference e-Reality: Constructing the e-Economy, Slovenia.

Osterwalder, A., & Pigneur, Y. (2010). *Business model generation: A handbook for visionaries, game changers, and challengers*. Hoboken, NJ: Wiley.

Osterwalder, A., & Pigneur, Y. (2013). Designing Business Models and Similar Strategic Objects: The Contribution of IS. *Journal of the Association for Information Systems, 14*(5), 237–244. doi:10.17705/1jais.00333

Plastrik, P., & Taylor, M. (2006). *Net Gains: A Handbook for Network Builders Seeking Social Change*. Retrieved from https://networkimpact.org/downloads/NetGainsHandbookVersion1.pdf

Prahalad, C. K., & Krishnan, M. S. (2008). *The New Age of Innovation: Driving Cocreated Value Through Global Networks*. McGraw-Hill.

Rappa, M. A. (2003). *Business Models on the Web, Managing the Digital Enterprise*. Retrieved from http://digitalenterprise.org

Rappa, M. A. (2004). The utility business model and the future of computing service. *IBM Systems Journal, 43*(1), 32–42. doi:10.1147j.431.0032

Rumelt, R. P. (1974). *Strategy.* Boston: Structure and Economic Performance.

Scott, B. R. (1973, March). Stages of Corporate Development. *Harvard Business Review.*

Shah, A. M. (2005). The Foundations of Successful Strategy Implementation. *Global Business Review, 6*(2), 293–302. doi:10.1177/097215090500600208

Tegmark, M. (2017). *Life 3.0: being human in the age of artificial intelligence.* Allen Lane.

Townsend, M., Coen, L., & Watson, K. (2017). From Regional to Global: Using a Network Strategy to Align a Multinational Organization. *People + Strategy, 40*(2).

Teece, D. J. (2010). Business Models, Business Strategy and Innovation. *Long Range Planning, 43*(2-3), 172–194. doi:10.1016/j.lrp.2009.07.003

Voelpel, S., Laibold, M., Tekie, E., & von Krogh, G. (2005). Escaping the Red Queen Effect in Competitive Strategy: Sense-testing Business Models. *European Management Journal, 23*(1), 37–49. doi:10.1016/j.emj.2004.12.008

deWit, B., & Meyer, R. (2010). *Strategy synthesis. Resolving strategy paradoxes to create competitive advantage* (3rd ed.). South-Western Cengage Learning.

ENDNOTES

[1]   Bayer CropScience is a leader in plant protection, biotechnology and sanitary hygiene. It holds a leading position in the market of insecticides and herbicides.

[2]   BOC Information Technologies Consulting GmbH was established by o. Univ. Prof. Dr. Dimitris Karagiannis (Website) in 1995 in Vienna, as a spin-off from the Business Process Management Systems (BPMS) Group from the department of Knowledge Engineering at the University of Vienna. Headquarters in Vienna and additional companies in Germany, Spain, Ireland, Greece, Poland, Switzerland and France, the BOC Group operates worldwide.

Conclusion

Management of contemporary organizations is a multi-faceted issue. It goes far beyond a search of excellence by optimizing intra-organizational processes. While the efficiency and effectiveness of organizational processes is an invariably desirable feature, management of the human factor and its potential is now coming to the fore. Human capital, initially in organizational and management theories, was treated as another, relatively expensive resource possessed by enterprises. Nowadays, in a knowledge-based economy, human capital is gaining in importance not only at the microeconomic but also at the macroeconomic level. Human is no longer solely a production factor that influences the implementation of intra-organizational processes and their excellence, but is becoming the potential that plays a more extensive role than that confined to an organization. Knowledge is the fundamental resource not only of a single organization but of entire economies. It is the foundation on which to develop innovations that are so strongly expected and desired in today's world.

At present, innovations are also changing their form. They are no longer closed innovations confined to a research laboratory or the R&D department of an enterprise. More and more often, they are open. Innovations are a response to expectations voiced by customers. This requires ongoing communication and exchange of views between organizations and their environment. This environment includes customers who express their expectations and needs. They are an important, but not the sole, element. Organizations today hold a permanent dialog not only with clients but also with partners, competitors, scientific institutes and administration. This list is not exhaustive because in an increasingly networked world, everyone is invited to debate and exchange opinions. And every opinion is valuable as it can provide an inspiration for further actions and innovations.

Technical infrastructure is critical for a dialog and exchange of information between parties. The development of ICT infrastructure has enabled the emergence and evolution of network structures that cross traditional organizational boundaries. Overcoming these institutional boundaries is a step towards innovation. This, in turn, is a key success factor for a modern organization that is expected to be ready to develop systematic and repeatable innovation. Being ready to cooperate within a network, organizations can use both physical and intellectual resources that are not available in the traditional system. The opportunity to attract representatives of various fields or cultures and to involve them in joint development of new solutions thereafter is a means to design a product that meets the expectations of the demanding market. The experience of companies such as Apple, IKEA or Tesla shows that a business model based on constant search of innovation is what contemporary economy expects from organizations. This economy is more and more commonly referred to as digital, given the ever more frequent digital transformation resulting from the development of technical infrastructure and more ICT solutions for data transmission, storage, processing and analysis. Digitization has become the sign of our times. It encompasses not only products and services but also, to an ever-larger extent, traditionally internal manufacturing processes (no one today is surprised by the possibility of designing their own sports shoes). Digital transformation is expanding to cover further areas of our lives. Not only products and services are available in their digital versions but also IT solutions are increasingly influencing communication methods. ICT infrastructure is a tool that supports learning or work. Under such conditions, a natural consequence is the development of cooperation networks that are put in place by means of ICT networks.

When the classical management scholars were formulating a theory describing organizations as open units that constantly exchanged resources or information with the environment, they certainly did not anticipate such intensive exchange as it is today. Regardless of whether we are dealing with a representative of business, public sector or science, it is necessary to establish relations and exchange various resources. The digital world in which we live is a world of continuous data exchange, and an environment where value (in any industry) is created through networks of links. All fields are being gradually digitalized, and even niche, traditional offline business activities (shoemaker, seamstress) – even if not consciously – resort to advice and recommendations of people active online.

The leading thesis of the monograph is the assumption that a network organization is a structure oriented towards knowledge management. We justify this thesis in individual chapters that simultaneously describe the various areas of organizational operations. Knowledge should be expanded in all such areas as evolution and strategic development of organizations and their structures, information technologies, innovation and decision-making. Today's organizations must be open to the inflow, uptake, development and subsequent exchange of knowledge with the environment. This is not an easy task. It requires enormous changes – structural ones, changes in culture, in operating methods, and in business models of organizations. There is no doubt, however, that this is a good course of action.

We are convinced that the world of connections, relations and knowledge exchange will exist in the next decades. What will the reality be like? We do not know. The world can be surprising, but that might be the reason why it is so immensely interesting. Where is management heading? Literature very often resorts to the following contrivance to answer the question "what next?". What already does or will exist is management 2.0 and then management 3.0. Sometimes a lateral trajectory of the management science development is mentioned, such as Industry 4.0. The idea by Max Tegmark, an MIT professor, founder and president of the Future of Life Institute is extremely interesting. He believes that forward-looking management involves knowledge used for AI (Artificial Intelligence) applications. This concept fits well in the issues of knowledge and innovation management in network organizations.

This monograph contributes to the discussion on the future of management sciences. It is characterized by openness, and it attempts to look at selected key problems of the future management processes. We will also be grateful to our readers for contact and any comments.

Related Readings

To continue IGI Global's long-standing tradition of advancing innovation through emerging research, please find below a compiled list of recommended IGI Global book chapters and journal articles in the areas of social enterprise, economic growth, and organizational development. These related readings will provide additional information and guidance to further enrich your knowledge and assist you with your own research.

Adams, S., Klobodu, E. K., & Lamptey, R. O. (2017). Electric Power Transmission, Distribution Losses, and Economic Growth in Ghana. In R. Das (Ed.), *Social, Health, and Environmental Infrastructures for Economic Growth* (pp. 216–233). Hershey, PA: IGI Global. doi:10.4018/978-1-5225-2364-2.ch011

Adams, S., Klobodu, E. K., & Lamptey, R. O. (2017). Health Infrastructure and Economic Growth in Sub-Saharan Africa. In R. Das (Ed.), *Social, Health, and Environmental Infrastructures for Economic Growth* (pp. 82–98). Hershey, PA: IGI Global. doi:10.4018/978-1-5225-2364-2.ch005

Adetiloye, K. A., Eke, P. O., & Taiwo, J. N. (2017). Public-Sector Project Abandonment Decision: A Test of the Ricardian Equivalence Theory on the Failed Lagos Metroline in Nigeria. In R. Das (Ed.), *Social, Health, and Environmental Infrastructures for Economic Growth* (pp. 188–215). Hershey, PA: IGI Global. doi:10.4018/978-1-5225-2364-2.ch010

Adhikary, M., & Khatun, M. (2016). Issues of Convergence: Some Evidences of SAARC Countries. In R. Das (Ed.), *Handbook of Research on Global Indicators of Economic and Political Convergence* (pp. 119–143). Hershey, PA: IGI Global. doi:10.4018/978-1-5225-0215-9.ch006

Akhter, T., & Palit, A. (2016). Effects of Neoliberal Globalization on Readymade Garment Industry of Bangladesh: A Case Study for Chittagong. In R. Das (Ed.), *Handbook of Research on Global Indicators of Economic and Political Convergence* (pp. 538–557). Hershey, PA: IGI Global. doi:10.4018/978-1-5225-0215-9.ch023

Akincilar, A. (2017). Do Judgmental Factors Lead to a Good Decision on Investing in a Currency or Mislead the Financial Player?: An Application in Turkey. In J. Stanković, P. Delias, S. Marinković, & S. Rochhia (Eds.), *Tools and Techniques for Economic Decision Analysis* (pp. 43–62). Hershey, PA: IGI Global. doi:10.4018/978-1-5225-0959-2.ch003

Allegreni, F. (2017). Crowdfunding as a Marketing Tool. In W. Vassallo (Ed.), *Crowdfunding for Sustainable Entrepreneurship and Innovation* (pp. 187–203). Hershey, PA: IGI Global. doi:10.4018/978-1-5225-0568-6.ch011

Altındağ, E. (2016). Current Approaches in Change Management. In A. Goksoy (Ed.), *Organizational Change Management Strategies in Modern Business* (pp. 24–51). Hershey, PA: IGI Global. doi:10.4018/978-1-4666-9533-7.ch002

Ambani, P. (2017). Crowdsourcing New Tools to Start Lean and Succeed in Entrepreneurship: Entrepreneurship in the Crowd Economy. In W. Vassallo (Ed.), *Crowdfunding for Sustainable Entrepreneurship and Innovation* (pp. 37–53). Hershey, PA: IGI Global. doi:10.4018/978-1-5225-0568-6.ch003

Amoako, G. K., Adjaison, G. K., & Osei-Bonsu, N. (2016). Role of Strategic Change Management in Emerging Markets: Ghanaian Perspective. In A. Goksoy (Ed.), *Organizational Change Management Strategies in Modern Business* (pp. 328–351). Hershey, PA: IGI Global. doi:10.4018/978-1-4666-9533-7.ch016

Arsenijević, O. M., Orčić, D., & Kastratović, E. (2017). Development of an Optimization Tool for Intangibles in SMEs: A Case Study from Serbia with a Pilot Research in the Prestige by Milka Company. In M. Vemić (Ed.), *Optimal Management Strategies in Small and Medium Enterprises* (pp. 320–347). Hershey, PA: IGI Global. doi:10.4018/978-1-5225-1949-2.ch015

Basu, I. (2017). Role of Infrastructure Development to Empower Women: An Over-Determined View. In R. Das (Ed.), *Social, Health, and Environmental Infrastructures for Economic Growth* (pp. 39–56). Hershey, PA: IGI Global. doi:10.4018/978-1-5225-2364-2.ch003

Ben Rejeb, W. (2017). Empirical Evidence on Corporate Governance Impact on CSR Disclosure in Developing Economies: The Tunisian and Egyptian Contexts. In D. Jamali (Ed.), *Comparative Perspectives on Global Corporate Social Responsibility* (pp. 116–137). Hershey, PA: IGI Global. doi:10.4018/978-1-5225-0720-8.ch006

Blomme, R. J., & Morsch, J. (2016). Organizations as Social Networks: The Role of the Compliance Officer as Agent of Change in Implementing Rules and Codes of Conduct. In A. Goksoy (Ed.), *Organizational Change Management Strategies in Modern Business* (pp. 110–121). Hershey, PA: IGI Global. doi:10.4018/978-1-4666-9533-7.ch006

Boachie, C. (2017). Decision Making under Risk and Uncertainty in the Oil and Gas Industry. In J. Stanković, P. Delias, S. Marinković, & S. Rochhia (Eds.), *Tools and Techniques for Economic Decision Analysis* (pp. 181–207). Hershey, PA: IGI Global. doi:10.4018/978-1-5225-0959-2.ch009

Candelise, C. (2017). The Application of Crowdfunding to the Energy Sector. In W. Vassallo (Ed.), *Crowdfunding for Sustainable Entrepreneurship and Innovation* (pp. 266–287). Hershey, PA: IGI Global. doi:10.4018/978-1-5225-0568-6.ch015

Carvalho, G., Simões, M., & Duarte, A. P. (2017). Export-Led Recovery in Portugal: Can It Also Sustain Growth? In J. Stanković, P. Delias, S. Marinković, & S. Rochhia (Eds.), *Tools and Techniques for Economic Decision Analysis* (pp. 129–150). Hershey, PA: IGI Global. doi:10.4018/978-1-5225-0959-2.ch006

Chakrabarti, G., & Sen, C. (2016). Green Convergence in Emerging Nations: The Determinants and the Possibilities. In R. Das (Ed.), *Handbook of Research on Global Indicators of Economic and Political Convergence* (pp. 448–473). Hershey, PA: IGI Global. doi:10.4018/978-1-5225-0215-9.ch020

Chakraborty, S. (2016). Economic Convergence and Real Dimensions: The Case of Shelter Deprivation. In R. Das (Ed.), *Handbook of Research on Global Indicators of Economic and Political Convergence* (pp. 338–363). Hershey, PA: IGI Global. doi:10.4018/978-1-5225-0215-9.ch015

Chan, D., & Kumar, S. (2017). Social Media Strategies for Small and Medium Scale Enterprise in the Klang Valley Region of Malaysia. In N. Ahmad, T. Ramayah, H. Halim, & S. Rahman (Eds.), *Handbook of Research on Small and Medium Enterprises in Developing Countries* (pp. 377–400). Hershey, PA: IGI Global. doi:10.4018/978-1-5225-2165-5.ch018

Chatterjee, N. (2017). An Assessment of Infrastructural Facilities in the Dryland Areas of West Bengal. In R. Das (Ed.), *Social, Health, and Environmental Infrastructures for Economic Growth* (pp. 234–268). Hershey, PA: IGI Global. doi:10.4018/978-1-5225-2364-2.ch012

Chatterjee, N., & Dinda, S. (2016). Convergence of Forest Resources in Jangalmahal, West Bengal. In R. Das (Ed.), *Handbook of Research on Global Indicators of Economic and Political Convergence* (pp. 511–537). Hershey, PA: IGI Global. doi:10.4018/978-1-5225-0215-9.ch022

Chatterjee, T., & Dinda, S. (2016). Health Status and Convergence in Developing Open Economies: Is Health Status Converging in Developing Economies? In R. Das (Ed.), *Handbook of Research on Global Indicators of Economic and Political Convergence* (pp. 364–381). Hershey, PA: IGI Global. doi:10.4018/978-1-5225-0215-9.ch016

Chaudhury, A. R. (2016). How Does Inequality of Opportunity Persist from Education to Earning?: Some Recent Evidences from India. In R. Das (Ed.), *Handbook of Research on Global Indicators of Economic and Political Convergence* (pp. 314–337). Hershey, PA: IGI Global. doi:10.4018/978-1-5225-0215-9.ch014

Cinelli, S. A. (2017). Real Estate Crowdfunding: 2015 and Beyond. In W. Vassallo (Ed.), *Crowdfunding for Sustainable Entrepreneurship and Innovation* (pp. 238–265). Hershey, PA: IGI Global. doi:10.4018/978-1-5225-0568-6.ch014

Cinelli, S. A. (2017). The World's Oldest Profession - Now and Then: Disruption of the Commercial Banking Model. In W. Vassallo (Ed.), *Crowdfunding for Sustainable Entrepreneurship and Innovation* (pp. 78–89). Hershey, PA: IGI Global. doi:10.4018/978-1-5225-0568-6.ch005

Çolakoğlu, S., Chung, Y., & Tarhan, A. B. (2016). Strategic Human Resource Management in Facilitating Organizational Change. In A. Goksoy (Ed.), *Organizational Change Management Strategies in Modern Business* (pp. 172–192). Hershey, PA: IGI Global. doi:10.4018/978-1-4666-9533-7.ch009

Damnjanović, A. M. (2017). Knowledge Management Optimization through IT and E-Business Utilization: A Qualitative Study on Serbian SMEs. In M. Vemić (Ed.), *Optimal Management Strategies in Small and Medium Enterprises* (pp. 249–267). Hershey, PA: IGI Global. doi:10.4018/978-1-5225-1949-2.ch012

Daneshpour, H. (2017). Integrating Sustainable Development into Project Portfolio Management through Application of Open Innovation. In M. Vemić (Ed.), *Optimal Management Strategies in Small and Medium Enterprises* (pp. 370–387). Hershey, PA: IGI Global. doi:10.4018/978-1-5225-1949-2.ch017

Das, R. C., Das, A., & Martin, F. (2016). Convergence Analysis of Households' Consumption Expenditure: A Cross Country Study. In R. Das (Ed.), *Handbook of Research on Global Indicators of Economic and Political Convergence* (pp. 1–28). Hershey, PA: IGI Global. doi:10.4018/978-1-5225-0215-9.ch001

Das, U., Das, R. C., & Ray, K. (2016). Convergence and Equality of Road Infrastructure: A Cross Country Analysis. In R. Das (Ed.), *Handbook of Research on Global Indicators of Economic and Political Convergence* (pp. 170–183). Hershey, PA: IGI Global. doi:10.4018/978-1-5225-0215-9.ch008

Dastaviz, A. H. (2016). Integration of Competing Values and Knowledge Organisational Activities in a New Model. *International Journal of Knowledge-Based Organizations*, 6(4), 50–61. doi:10.4018/IJKBO.2016100105

Datta, P. K., Tiwary, H., & Chakrabarti, S. (2017). MGNREGS of India: Complementarities Between Employment and Infrastructure. In R. Das (Ed.), *Social, Health, and Environmental Infrastructures for Economic Growth* (pp. 57–81). Hershey, PA: IGI Global. doi:10.4018/978-1-5225-2364-2.ch004

Davidson, D. K., & Yin, J. (2017). Corporate Social Responsibility (CSR) in China: A Contextual Exploration. In D. Jamali (Ed.), *Comparative Perspectives on Global Corporate Social Responsibility* (pp. 28–48). Hershey, PA: IGI Global. doi:10.4018/978-1-5225-0720-8.ch002

de Burgh-Woodman, H., Bressan, A., & Torrisi, A. (2017). An Evaluation of the State of the CSR Field in Australia: Perspectives from the Banking and Mining Sectors. In D. Jamali (Ed.), *Comparative Perspectives on Global Corporate Social Responsibility* (pp. 138–164). Hershey, PA: IGI Global. doi:10.4018/978-1-5225-0720-8.ch007

De Ruiter, M., Blomme, R. J., & Schalk, R. (2016). Reducing the Negative Effects of Psychological Contract Breach during Management-Imposed Change: A Trickle-Down Model of Management Practices. In A. Goksoy (Ed.), *Organizational Change Management Strategies in Modern Business* (pp. 122–142). Hershey, PA: IGI Global. doi:10.4018/978-1-4666-9533-7.ch007

Dedousis, E., & Rutter, R. N. (2016). Workforce Localisation and Change Management: The View from the Gulf. In A. Goksoy (Ed.), *Organizational Change Management Strategies in Modern Business* (pp. 301–327). Hershey, PA: IGI Global. doi:10.4018/978-1-4666-9533-7.ch015

Demiray, M., & Aslanbay, Y. (2017). The Crowdfunding Communities and the Value of Identification for Sustainability of Co-Creation. In W. Vassallo (Ed.), *Crowdfunding for Sustainable Entrepreneurship and Innovation* (pp. 155–174). Hershey, PA: IGI Global. doi:10.4018/978-1-5225-0568-6.ch009

Demiray, M., Burnaz, S., & Aslanbay, Y. (2017). The Crowdfunding Market, Models, Platforms, and Projects. In W. Vassallo (Ed.), *Crowdfunding for Sustainable Entrepreneurship and Innovation* (pp. 90–126). Hershey, PA: IGI Global. doi:10.4018/978-1-5225-0568-6.ch006

Dikeç, A., Kane, V., & Çapar, N. (2017). Cross-Country and Cross-Sector CSR Variations: A Comparative Analysis of CSR Reporting in the U.S., South Korea, and Turkey. In D. Jamali (Ed.), *Comparative Perspectives on Global Corporate Social Responsibility* (pp. 69–95). Hershey, PA: IGI Global. doi:10.4018/978-1-5225-0720-8.ch004

Duruji, M. M., & Urenma, D. F. (2016). The Environmentalism and Politics of Climate Change: A Study of the Process of Global Convergence through UNFCCC Conferences. In R. Das (Ed.), *Handbook of Research on Global Indicators of Economic and Political Convergence* (pp. 398-429). Hershey, PA: IGI Global. 10.4018/978-1-5225-0215-9.ch018

Erne, R. (2016). Change Management Revised. In A. Goksoy (Ed.), *Organizational Change Management Strategies in Modern Business* (pp. 1–23). Hershey, PA: IGI Global. doi:10.4018/978-1-4666-9533-7.ch001

Ertürk, E., Yılmaz, D., & Çetin, I. (2016). Optimum Currency Area Theory and Business Cycle Convergence in EMU: Considering the Sovereign Debt Crisis. In R. Das (Ed.), *Handbook of Research on Global Indicators of Economic and Political Convergence* (pp. 67–91). Hershey, PA: IGI Global. doi:10.4018/978-1-5225-0215-9.ch004

Eryılmaz, M. E., & Eryılmaz, F. (2016). Change Emphasis in Mission and Vision Statements of the First 1000 Turkish Organizations: A Content Analysis. In A. Goksoy (Ed.), *Organizational Change Management Strategies in Modern Business* (pp. 352–362). Hershey, PA: IGI Global. doi:10.4018/978-1-4666-9533-7.ch017

Frączkiewicz-Wronka, A., Tkacz, M., & Arando, S. (2017). The Business Model of a Public Social Partnership: Contextual Determinants. In M. Lewandowski & B. Kożuch (Eds.), *Public Sector Entrepreneurship and the Integration of Innovative Business Models* (pp. 108–138). Hershey, PA: IGI Global. doi:10.4018/978-1-5225-2215-7.ch005

Friedrich, P., & Chebotareva, M. (2017). Options for Applying Functional Overlapping Competing Jurisdictions (FOCJs) for Municipal Cooperation in Russia. In M. Lewandowski & B. Kożuch (Eds.), *Public Sector Entrepreneurship and the Integration of Innovative Business Models* (pp. 73–107). Hershey, PA: IGI Global. doi:10.4018/978-1-5225-2215-7.ch004

Gan, P. L., Mahmud, I., Ramayah, T., & Zuhora, F. T. (2017). Understanding Attitude towards Green IT among Professionals in IT Service SMEs in Bangladesh. In N. Ahmad, T. Ramayah, H. Halim, & S. Rahman (Eds.), *Handbook of Research on Small and Medium Enterprises in Developing Countries* (pp. 48–66). Hershey, PA: IGI Global. doi:10.4018/978-1-5225-2165-5.ch003

Ganesan, Y., Haron, H., Amran, A., & Ooi, S. K. (2017). Corporate Social Responsibility in SMEs: The Role of Non-Audit Services. In N. Ahmad, T. Ramayah, H. Halim, & S. Rahman (Eds.), *Handbook of Research on Small and Medium Enterprises in Developing Countries* (pp. 345–359). Hershey, PA: IGI Global. doi:10.4018/978-1-5225-2165-5.ch016

Ghosh, D., & Dinda, S. (2017). Health Infrastructure and Economic Development in India. In R. Das (Ed.), *Social, Health, and Environmental Infrastructures for Economic Growth* (pp. 99–119). Hershey, PA: IGI Global. doi:10.4018/978-1-5225-2364-2.ch006

Ghosh, S. (2017). Changes in Infrastructural Condition of Slums in India: A State Level Analysis. In R. Das (Ed.), *Social, Health, and Environmental Infrastructures for Economic Growth* (pp. 14–38). Hershey, PA: IGI Global. doi:10.4018/978-1-5225-2364-2.ch002

Gonzalez, R. V. (2016). Knowledge Management Process in Multi-Site Provision of Service. *International Journal of Knowledge Management*, *12*(2), 20–37. doi:10.4018/IJKM.2016040102

Greco, T. (2017). Horizontal Crowdfunding Platforms: Socio-Economic Impact and Exogenous/Endogenous System of Rules. In W. Vassallo (Ed.), *Crowdfunding for Sustainable Entrepreneurship and Innovation* (pp. 127–139). Hershey, PA: IGI Global. doi:10.4018/978-1-5225-0568-6.ch007

Greco, T. (2017). Vertical and Community-Based Crowdfunding Platforms. In W. Vassallo (Ed.), *Crowdfunding for Sustainable Entrepreneurship and Innovation* (pp. 140–154). Hershey, PA: IGI Global. doi:10.4018/978-1-5225-0568-6.ch008

Hack-Polay, D., & Qiu, H. (2017). Doing Good Doing Well: Discussion of CSR in the Pulp and Paper Industry in the Asian Context. In D. Jamali (Ed.), *Comparative Perspectives on Global Corporate Social Responsibility* (pp. 226–240). Hershey, PA: IGI Global. doi:10.4018/978-1-5225-0720-8.ch011

Halim, H. A., Ahmad, N. H., Hanifah, H., & Ramayah, T. (2017). The Impediments to Entrepreneurial Ventures among the Bottom of Pyramid Community in Northern Malaysia. In N. Ahmad, T. Ramayah, H. Halim, & S. Rahman (Eds.), *Handbook of Research on Small and Medium Enterprises in Developing Countries* (pp. 67–81). Hershey, PA: IGI Global. doi:10.4018/978-1-5225-2165-5.ch004

Hanifah, H., Halim, H. A., Ahmad, N. H., & Vafaei-Zadeh, A. (2017). Innovation Culture as a Mediator Between Specific Human Capital and Innovation Performance Among Bumiputera SMEs in Malaysia. In N. Ahmad, T. Ramayah, H. Halim, & S. Rahman (Eds.), *Handbook of Research on Small and Medium Enterprises in Developing Countries* (pp. 261–279). Hershey, PA: IGI Global. doi:10.4018/978-1-5225-2165-5.ch012

Haro-de-Rosario, A., del Mar Gálvez-Rodríguez, M., & Caba-Pérez, M. D. (2017). Determinants of Corporate Social Responsibility Disclosure in Latin American Companies: An Analysis of the Oil and Gas Sector. In D. Jamali (Ed.), *Comparative Perspectives on Global Corporate Social Responsibility* (pp. 165–184). Hershey, PA: IGI Global. doi:10.4018/978-1-5225-0720-8.ch008

Hassaan, M. (2016). Sustainability Reporting in Transitional Economies. In R. Das (Ed.), *Handbook of Research on Global Indicators of Economic and Political Convergence* (pp. 184–204). Hershey, PA: IGI Global. doi:10.4018/978-1-5225-0215-9.ch009

Hassan, A., & Lund-Thomsen, P. (2017). Multi-Stakeholder Initiatives and Corporate Social Responsibility in Global Value Chains: Towards an Analytical Framework and a Methodology. In D. Jamali (Ed.), *Comparative Perspectives on Global Corporate Social Responsibility* (pp. 241–257). Hershey, PA: IGI Global. doi:10.4018/978-1-5225-0720-8.ch012

Hossain, M. M., & Ibrahim, Y. (2017). Working Capital Financing by Banks in Small Enterprises: Problems and Challenges for Bangladesh. In N. Ahmad, T. Ramayah, H. Halim, & S. Rahman (Eds.), *Handbook of Research on Small and Medium Enterprises in Developing Countries* (pp. 137–158). Hershey, PA: IGI Global. doi:10.4018/978-1-5225-2165-5.ch007

Indiran, L., Khalifah, Z., Ismail, K., & Ramanathan, S. (2017). Business Incubation in Malaysia: An Overview of Multimedia Super Corridor, Small and Medium Enterprises, and Incubators in Malaysia. In N. Ahmad, T. Ramayah, H. Halim, & S. Rahman (Eds.), *Handbook of Research on Small and Medium Enterprises in Developing Countries* (pp. 322–344). Hershey, PA: IGI Global. doi:10.4018/978-1-5225-2165-5.ch015

Issa, T., & Pick, D. (2017). Teaching Business Ethics Post GFC: A Corporate Social Responsibility of Universities. In D. Jamali (Ed.), *Comparative Perspectives on Global Corporate Social Responsibility* (pp. 290–307). Hershey, PA: IGI Global. doi:10.4018/978-1-5225-0720-8.ch015

Ivana, B., & Vrkić, F. (2017). Crisis Communication and Crisis Management during the Crisis: Case Study of Croatia. In J. Stanković, P. Delias, S. Marinković, & S. Rochhia (Eds.), *Tools and Techniques for Economic Decision Analysis* (pp. 208–224). Hershey, PA: IGI Global. doi:10.4018/978-1-5225-0959-2.ch010

Ivančević, K. (2017). Insurance as an Optimization Tool for Risk Management in Small and Medium-Sized Enterprises. In M. Vemić (Ed.), *Optimal Management Strategies in Small and Medium Enterprises* (pp. 178–197). Hershey, PA: IGI Global. doi:10.4018/978-1-5225-1949-2.ch009

Jain, P. (2017). A Crowd-Funder Value (CFV) Framework for Crowd-Investment: A Roadmap for Entrepreneurial Success in the Contemporary Society. In W. Vassallo (Ed.), *Crowdfunding for Sustainable Entrepreneurship and Innovation* (pp. 288–309). Hershey, PA: IGI Global. doi:10.4018/978-1-5225-0568-6.ch016

Jalali, A., Thurasamy, R., & Jaafar, M. (2017). The Moderating Effect of Social Capital in Relation to Entrepreneurial Orientation and Firm Performance. In N. Ahmad, T. Ramayah, H. Halim, & S. Rahman (Eds.), *Handbook of Research on Small and Medium Enterprises in Developing Countries* (pp. 82–115). Hershey, PA: IGI Global. doi:10.4018/978-1-5225-2165-5.ch005

Jana, S. K., & Karmakar, A. K. (2016). Food Security in Asia: Is There Convergence? In R. Das (Ed.), *Handbook of Research on Global Indicators of Economic and Political Convergence* (pp. 382–397). Hershey, PA: IGI Global. doi:10.4018/978-1-5225-0215-9.ch017

Jana, S. K., & Karmakar, A. K. (2017). Infrastructure, Education, and Economic Development in India: A State Level Analysis. In R. Das (Ed.), *Social, Health, and Environmental Infrastructures for Economic Growth* (pp. 1–13). Hershey, PA: IGI Global. doi:10.4018/978-1-5225-2364-2.ch001

Jankovic-Milic, V., & Džunić, M. (2017). Measuring Governance: The Application of Grey Relational Analysis on World Governance Indicators. In J. Stanković, P. Delias, S. Marinković, & S. Rochhia (Eds.), *Tools and Techniques for Economic Decision Analysis* (pp. 104–128). Hershey, PA: IGI Global. doi:10.4018/978-1-5225-0959-2.ch005

Jayapal, P., & Omar, A. (2017). The Role of Value Co-Creation on Brand Image: A Conceptual Framework for the Market Performance of SMEs in Malaysia. In N. Ahmad, T. Ramayah, H. Halim, & S. Rahman (Eds.), *Handbook of Research on Small and Medium Enterprises in Developing Countries* (pp. 185–207). Hershey, PA: IGI Global. doi:10.4018/978-1-5225-2165-5.ch009

Kaplan, J., & Montiel, I. (2017). East vs. West Approaches to Reporting Corporate Sustainability Strategies to the World: Corporate Sustainability Reporting: East vs. West. In D. Jamali (Ed.), *Comparative Perspectives on Global Corporate Social Responsibility* (pp. 49–68). Hershey, PA: IGI Global. doi:10.4018/978-1-5225-0720-8.ch003

Kasemsap, K. (2016). The Roles of Organizational Change Management and Resistance to Change in the Modern Business World. In A. Goksoy (Ed.), *Organizational Change Management Strategies in Modern Business* (pp. 143–171). Hershey, PA: IGI Global. doi:10.4018/978-1-4666-9533-7.ch008

Kasemsap, K. (2017). Encouraging Internationalization and Entrepreneurial Orientation in Small and Medium Enterprises. In N. Ahmad, T. Ramayah, H. Halim, & S. Rahman (Eds.), *Handbook of Research on Small and Medium Enterprises in Developing Countries* (pp. 233–259). Hershey, PA: IGI Global. doi:10.4018/978-1-5225-2165-5.ch011

Kesti, M. O., Leinonen, J., & Kesti, T. (2017). The Productive Leadership Game: From Theory to Game-Based Learning. In M. Lewandowski & B. Kożuch (Eds.), *Public Sector Entrepreneurship and the Integration of Innovative Business Models* (pp. 238–260). Hershey, PA: IGI Global. doi:10.4018/978-1-5225-2215-7.ch010

King, D. R. (2016). Management as a Limit to Organizational Change: Implications for Acquisitions. In A. Goksoy (Ed.), *Organizational Change Management Strategies in Modern Business* (pp. 52–73). Hershey, PA: IGI Global. doi:10.4018/978-1-4666-9533-7.ch003

Kishna, T., Blomme, R. J., & van der Veen, J. A. (2016). Organizational Routines: Developing a Duality Model to Explain the Effects of Strategic Change Initiatives. In A. Goksoy (Ed.), *Organizational Change Management Strategies in Modern Business* (pp. 363–385). Hershey, PA: IGI Global. doi:10.4018/978-1-4666-9533-7.ch018

Kleverlaan, R. (2017). How to Launch a Successful Crowdfunding Campaign. In W. Vassallo (Ed.), *Crowdfunding for Sustainable Entrepreneurship and Innovation* (pp. 224–237). Hershey, PA: IGI Global. doi:10.4018/978-1-5225-0568-6.ch013

Kożuch, B., & Jabłoński, A. (2017). Adopting the Concept of Business Models in Public Management. In M. Lewandowski & B. Kożuch (Eds.), *Public Sector Entrepreneurship and the Integration of Innovative Business Models* (pp. 10–46). Hershey, PA: IGI Global. doi:10.4018/978-1-5225-2215-7.ch002

Laha, A. (2016). Association between Governance and Human Development in South Asia: A Cross Country Analysis. In R. Das (Ed.), *Handbook of Research on Global Indicators of Economic and Political Convergence* (pp. 254–273). Hershey, PA: IGI Global. doi:10.4018/978-1-5225-0215-9.ch012

Lavassani, K. M., & Movahedi, B. (2017). Applications Driven Information Systems: Beyond Networks toward Business Ecosystems. *International Journal of Innovation in the Digital Economy*, 8(1), 61–75. doi:10.4018/IJIDE.2017010104

Lerro, A. M. (2017). Legal Aspects and Regulation in Crowdfunding: Comparisons across Countries. In W. Vassallo (Ed.), *Crowdfunding for Sustainable Entrepreneurship and Innovation* (pp. 204–223). Hershey, PA: IGI Global. doi:10.4018/978-1-5225-0568-6.ch012

Lewandowski, M. (2017). Public Organizations and Business Model Innovation: The Role of Public Service Design. In M. Lewandowski & B. Kożuch (Eds.), *Public Sector Entrepreneurship and the Integration of Innovative Business Models* (pp. 47–72). Hershey, PA: IGI Global. doi:10.4018/978-1-5225-2215-7.ch003

Lousã, M. D., & Monteiro, J. A. (2015). Explanatory Model of Adoption, Development and Utilization of Administrative Workflow Systems. *International Journal of Sociotechnology and Knowledge Development*, 7(4), 31–52. doi:10.4018/IJSKD.2015100103

Maciejczak, M., & Słodki, A. (2017). Initial Price Strategies of Polish Micro and Small Enterprises: An Application of Game Theory for Industrial Organization of the SME Sector. In M. Vemić (Ed.), *Optimal Management Strategies in Small and Medium Enterprises* (pp. 126–143). Hershey, PA: IGI Global. doi:10.4018/978-1-5225-1949-2.ch007

Magala, S. J. (2017). Between Davos and Porto Alegre: Democratic Entrepreneurship as Crowdsourcing for Ideas. In M. Lewandowski & B. Kożuch (Eds.), *Public Sector Entrepreneurship and the Integration of Innovative Business Models* (pp. 1–9). Hershey, PA: IGI Global. doi:10.4018/978-1-5225-2215-7.ch001

Malega, P. (2017). Small and Medium Enterprises in the Slovak Republic: Status and Competitiveness of SMEs in the Global Markets and Possibilities of Optimization. In M. Vemić (Ed.), *Optimal Management Strategies in Small and Medium Enterprises* (pp. 102–124). Hershey, PA: IGI Global. doi:10.4018/978-1-5225-1949-2.ch006

Malek, N. A., & Takala, J. A. (2017). Regression Analysis for Environmental Practices: Participation among Finnish SMEs. In M. Vemić (Ed.), *Optimal Management Strategies in Small and Medium Enterprises* (pp. 44–59). Hershey, PA: IGI Global. doi:10.4018/978-1-5225-1949-2.ch003

Malik, A. (2016). The Role of HR Strategies in Change. In A. Goksoy (Ed.), *Organizational Change Management Strategies in Modern Business* (pp. 193–215). Hershey, PA: IGI Global. doi:10.4018/978-1-4666-9533-7.ch010

Mandal, C., & Gupta, A. C. (2016). A Comparative Study on World-Wide Carbon Emission Convergence: An Empirical Analysis. In R. Das (Ed.), *Handbook of Research on Global Indicators of Economic and Political Convergence* (pp. 430–447). Hershey, PA: IGI Global. doi:10.4018/978-1-5225-0215-9.ch019

Manzoor, A. (2017). Sustainable Infrastructures: A New Infrastructure Investment Strategy. In R. Das (Ed.), *Social, Health, and Environmental Infrastructures for Economic Growth* (pp. 145–164). Hershey, PA: IGI Global. doi:10.4018/978-1-5225-2364-2.ch008

Mazumdar, D. (2016). The Problems of Development Gap between Developed and Developing Nations: Is There Any Sign of Convergence? In R. Das (Ed.), *Handbook of Research on Global Indicators of Economic and Political Convergence* (pp. 29–50). Hershey, PA: IGI Global. doi:10.4018/978-1-5225-0215-9.ch002

Milošević, O. (2017). Critical Review of SME Regulation Optimization in Serbia: A Reflection on Harmonization with the EU Acquis. In M. Vemić (Ed.), *Optimal Management Strategies in Small and Medium Enterprises* (pp. 79–101). Hershey, PA: IGI Global. doi:10.4018/978-1-5225-1949-2.ch005

Molnar, R. M. (2017). Structural Change Management for Sustainable SME Development: Applying Classical Management Tools. In M. Vemić (Ed.), *Optimal Management Strategies in Small and Medium Enterprises* (pp. 348–369). Hershey, PA: IGI Global. doi:10.4018/978-1-5225-1949-2.ch016

Monahov, A. (2017). The Effects of Prudential Supervision on Bank Resiliency and Profits in a Multi-Agent Setting. In J. Stanković, P. Delias, S. Marinković, & S. Rochhia (Eds.), *Tools and Techniques for Economic Decision Analysis* (pp. 63–103). Hershey, PA: IGI Global. doi:10.4018/978-1-5225-0959-2.ch004

Mukherjee, S. (2017). Anatomy and Significance of Public Healthcare Expenditure and Economic Growth Nexus in India: Its Implications for Public Health Infrastructure Thereof. In R. Das (Ed.), *Social, Health, and Environmental Infrastructures for Economic Growth* (pp. 120–144). Hershey, PA: IGI Global. doi:10.4018/978-1-5225-2364-2.ch007

Mukhopadhyay, D., & Mandal, A. K. (2017). Infrastructure Development and Changing Market Penetration of Consumer Durables in Rural India: An Empirical Investigation. In R. Das (Ed.), *Social, Health, and Environmental Infrastructures for Economic Growth* (pp. 269–289). Hershey, PA: IGI Global. doi:10.4018/978-1-5225-2364-2.ch013

Mukhopadhyay, P., Sinha, M., & Sengupta, P. P. (2017). Importance of Sustainable Rural Development through Agrarian Reforms: An Indian Scenario. In R. Das (Ed.), *Social, Health, and Environmental Infrastructures for Economic Growth* (pp. 290–306). Hershey, PA: IGI Global. doi:10.4018/978-1-5225-2364-2.ch014

Musa, E. (2017). Crowdsourcing Social Innovation: Towards a Collaborative Social Capitalism. In W. Vassallo (Ed.), *Crowdfunding for Sustainable Entrepreneurship and Innovation* (pp. 54–77). Hershey, PA: IGI Global. doi:10.4018/978-1-5225-0568-6.ch004

Nekaj, E. L. (2017). The Crowd Economy: From the Crowd to Businesses to Public Administrations and Multinational Companies. In W. Vassallo (Ed.), *Crowdfunding for Sustainable Entrepreneurship and Innovation* (pp. 1–19). Hershey, PA: IGI Global. doi:10.4018/978-1-5225-0568-6.ch001

Neogi, C. (2016). Productivity Convergence and Asian Trade Blocks. In R. Das (Ed.), *Handbook of Research on Global Indicators of Economic and Political Convergence* (pp. 92–118). Hershey, PA: IGI Global. doi:10.4018/978-1-5225-0215-9.ch005

Niederer, P. M. (2017). Principles in Crowdfunding Benefits and Validation Tools: How to Ensure Campaign Success. In W. Vassallo (Ed.), *Crowdfunding for Sustainable Entrepreneurship and Innovation* (pp. 175–186). Hershey, PA: IGI Global. doi:10.4018/978-1-5225-0568-6.ch010

Omiunu, O. G. (2017). Deploying ICT to Enhance Small Businesses and Achieving Sustainable Development: A Paradigm to Reducing Poverty and Unemployment and Enhancing a Sustainable National Development in Nigeria. In N. Ahmad, T. Ramayah, H. Halim, & S. Rahman (Eds.), *Handbook of Research on Small and Medium Enterprises in Developing Countries* (pp. 208–232). Hershey, PA: IGI Global. doi:10.4018/978-1-5225-2165-5.ch010

Osman, M. N. (2017). Internet-Based Social Reporting in Emerging Economies: Insights from Public Banks in Egypt and the UAE. In D. Jamali (Ed.), *Comparative Perspectives on Global Corporate Social Responsibility* (pp. 96–115). Hershey, PA: IGI Global. doi:10.4018/978-1-5225-0720-8.ch005

Özgeldi, M. (2016). Role of Human Resources in Change. In A. Goksoy (Ed.), *Organizational Change Management Strategies in Modern Business* (pp. 216–229). Hershey, PA: IGI Global. doi:10.4018/978-1-4666-9533-7.ch011

Patra, S. (2016). Role of SAARC in Convergence of South Asian Economies. In R. Das (Ed.), *Handbook of Research on Global Indicators of Economic and Political Convergence* (pp. 144–169). Hershey, PA: IGI Global. doi:10.4018/978-1-5225-0215-9.ch007

Patra, S. (2017). Energy Challenges and Infrastructure Development in South Asia. In R. Das (Ed.), *Social, Health, and Environmental Infrastructures for Economic Growth* (pp. 165–187). Hershey, PA: IGI Global. doi:10.4018/978-1-5225-2364-2.ch009

Puaschunder, J. (2017). The Call for Global Responsible Inter-Generational Leadership: The Quest of an Integration of Inter-Generational Equity in Corporate Social Responsibility (CSR) Models. In D. Jamali (Ed.), *Comparative Perspectives on Global Corporate Social Responsibility* (pp. 276–289). Hershey, PA: IGI Global. doi:10.4018/978-1-5225-0720-8.ch014

Radosavljevic, M., & Andjelkovic, A. (2017). Multi-Criteria Decision Making Approach for Choosing Business Process for the Improvement: Upgrading of the Six Sigma Methodology. In J. Stanković, P. Delias, S. Marinković, & S. Rochhia (Eds.), *Tools and Techniques for Economic Decision Analysis* (pp. 225–247). Hershey, PA: IGI Global. doi:10.4018/978-1-5225-0959-2.ch011

Rahman, S. A., Ahmad, N. H., & Taghizadeh, S. K. (2017). On the Road to SME Sector Development in Bangladesh: A Guideline Based on Current Challenges and Opportunities. In N. Ahmad, T. Ramayah, H. Halim, & S. Rahman (Eds.), *Handbook of Research on Small and Medium Enterprises in Developing Countries* (pp. 117–136). Hershey, PA: IGI Global. doi:10.4018/978-1-5225-2165-5.ch006

Raimi, L. (2017). Leveraging CSR as a 'support-aid' for Triple Bottom-Line Development in Nigeria: Evidence from the Telecommunication Industry. In D. Jamali (Ed.), *Comparative Perspectives on Global Corporate Social Responsibility* (pp. 208–225). Hershey, PA: IGI Global. doi:10.4018/978-1-5225-0720-8.ch010

Rajamanickam, S. (2016). Exploring Landscapes in Regional Convergence: Environment and Sustainable Development in South Asia. In R. Das (Ed.), *Handbook of Research on Global Indicators of Economic and Political Convergence* (pp. 474–510). Hershey, PA: IGI Global. doi:10.4018/978-1-5225-0215-9.ch021

Ray, K., Das, R. C., & Das, U. (2016). Convergence Aspect of Capital Formation: A Study on Major Countries. In R. Das (Ed.), *Handbook of Research on Global Indicators of Economic and Political Convergence* (pp. 51–66). Hershey, PA: IGI Global. doi:10.4018/978-1-5225-0215-9.ch003

Richet, J. (2016). Internal Communication Failure in Times of Change. In A. Goksoy (Ed.), *Organizational Change Management Strategies in Modern Business* (pp. 289–300). Hershey, PA: IGI Global. doi:10.4018/978-1-4666-9533-7.ch014

Riemann, U. (2016). The Power of Three: A Blended Approach of Project-, Change Management, and Design Thinking. In A. Goksoy (Ed.), *Organizational Change Management Strategies in Modern Business* (pp. 74–94). Hershey, PA: IGI Global. doi:10.4018/978-1-4666-9533-7.ch004

Rossetti di Valdalbero, D., & Birnbaum, B. (2017). Towards a New Economy: Co-Creation and Open Innovation in a Trustworthy Europe. In W. Vassallo (Ed.), *Crowdfunding for Sustainable Entrepreneurship and Innovation* (pp. 20–36). Hershey, PA: IGI Global. doi:10.4018/978-1-5225-0568-6.ch002

Saglietto, L., David, D., & Cezanne, C. (2017). Rethinking Social Capital Measurement. In J. Stanković, P. Delias, S. Marinković, & S. Rochhia (Eds.), *Tools and Techniques for Economic Decision Analysis* (pp. 248–268). Hershey, PA: IGI Global. doi:10.4018/978-1-5225-0959-2.ch012

Sanchez-Ruiz, L., & Blanco, B. (2017). Process Management for SMEs: Barriers, Enablers, and Benefits. In M. Vemić (Ed.), *Optimal Management Strategies in Small and Medium Enterprises* (pp. 293–319). Hershey, PA: IGI Global. doi:10.4018/978-1-5225-1949-2.ch014

Savic, M., & Zubovic, J. (2017). Modeling Labor Market Flows on the Basis of Sectoral Employment in Europe. In J. Stanković, P. Delias, S. Marinković, & S. Rochhia (Eds.), *Tools and Techniques for Economic Decision Analysis* (pp. 151–169). Hershey, PA: IGI Global. doi:10.4018/978-1-5225-0959-2.ch007

Scholtz, B., Koorsse, M., & Loleka, S. (2017). Understanding and Adoption of E-Finance in Small and Medium Enterprises (SMEs) in Developing Countries: A Study of Bangladesh and South Africa. In N. Ahmad, T. Ramayah, H Halim, & S. Rahman (Eds.), *Handbook of Research on Small and Medium Enterprises in Developing Countries* (pp. 159–184). Hershey, PA: IGI Global. doi:10.4018/978-1-5225-2165-5.ch008

Šebestová, J., & Palová, Z. (2017). Support of Social Innovations: Case of the Czech Republic. In M. Lewandowski & B. Kożuch (Eds.), *Public Sector Entrepreneurship and the Integration of Innovative Business Models* (pp. 165–187). Hershey, PA: IGI Global. doi:10.4018/978-1-5225-2215-7.ch007

Seliga, R. (2017). Public Sector Marketing in Poland. In M. Lewandowski & B. Kożuch (Eds.), *Public Sector Entrepreneurship and the Integration of Innovative Business Models* (pp. 214–237). Hershey, PA: IGI Global. doi:10.4018/978-1-5225-2215-7.ch009

Šević, A., & Marinković, S. (2017). Investment Decision Making: Where Do We Stand? In J. Stanković, P. Delias, S. Marinković, & S. Rochhia (Eds.), *Tools and Techniques for Economic Decision Analysis* (pp. 1–23). Hershey, PA: IGI Global. doi:10.4018/978-1-5225-0959-2.ch001

Şimşit, Z. T., Günay, N. S., & Vayvay, Ö. (2016). Organizational Learning to Managing Change: Key Player of Continuous Improvement in the 21st Century. In A. Goksoy (Ed.), *Organizational Change Management Strategies in Modern Business* (pp. 95–109). Hershey, PA: IGI Global. doi:10.4018/978-1-4666-9533-7.ch005

Širůček, M., & Křen, L. (2017). Application of Markowitz Portfolio Theory by Building Optimal Portfolio on the US Stock Market. In J. Stanković, P. Delias, S. Marinković, & S. Rochhia (Eds.), *Tools and Techniques for Economic Decision Analysis* (pp. 24–42). Hershey, PA: IGI Global. doi:10.4018/978-1-5225-0959-2.ch002

Soniewicki, M. (2017). The Importance of Market Orientation in Creating a Competitive Advantage of Micro, Small, and Medium-Sized Companies in the Internationalization Process. In M. Vemić (Ed.), *Optimal Management Strategies in Small and Medium Enterprises* (pp. 1–21). Hershey, PA: IGI Global. doi:10.4018/978-1-5225-1949-2.ch001

Staszewska, B. M. (2017). Local Public Enterprise Business Model as Multiple Value Creation System. In M. Lewandowski & B. Kożuch (Eds.), *Public Sector Entrepreneurship and the Integration of Innovative Business Models* (pp. 188–213). Hershey, PA: IGI Global. doi:10.4018/978-1-5225-2215-7.ch008

Stefanova, J. S., & Wenner, Z. (2017). Entrepreneurship and Enterprise Value Creation in Support of Smart, Sustainable, and Inclusive Growth in the European Union. In M. Vemić (Ed.), *Optimal Management Strategies in Small and Medium Enterprises* (pp. 268–292). Hershey, PA: IGI Global. doi:10.4018/978-1-5225-1949-2.ch013

Taghizadeh, S. K., Rahman, S. A., & Ramayah, T. (2017). Innovation-Driven Planned Behaviour Towards Achieving the Wellbeing of the Malaysian SMEs. In N. Ahmad, T. Ramayah, H. Halim, & S. Rahman (Eds.), *Handbook of Research on Small and Medium Enterprises in Developing Countries* (pp. 280–296). Hershey, PA: IGI Global. doi:10.4018/978-1-5225-2165-5.ch013

Thavinpipatkul, C., Ratana-Ubol, A., & Charungkaittikul, S. (2016). Transformative Learning Factors to Enhance Integral Healthy Organizations. *International Journal of Adult Vocational Education and Technology*, 7(1), 65–83. doi:10.4018/IJAVET.2016010105

Theodoridis, A., Ragkos, A., Angelidis, P., Batzios, C., & Samathrakis, V. (2017). Typologies as Management Tools: Understanding the Environmental Attitudes and Economic Prospects of Mussel Farmers in Greece. In J. Stanković, P. Delias, S. Marinković, & S. Rochhia (Eds.), *Tools and Techniques for Economic Decision Analysis* (pp. 170–180). Hershey, PA: IGI Global. doi:10.4018/978-1-5225-0959-2.ch008

Thongpoon, S., Ahmad, N. H., & Mahmud, I. (2017). Sustainable Approach towards Thai SMEs: The Effects of Country Philosophy of Sufficiency Economy and Government Support. In N. Ahmad, T. Ramayah, H. Halim, & S. Rahman (Eds.), *Handbook of Research on Small and Medium Enterprises in Developing Countries* (pp. 1–22). Hershey, PA: IGI Global. doi:10.4018/978-1-5225-2165-5.ch001

Timilsina, B. (2017). Overcoming the Barriers of Strategic Planning, Implementation, and Monitoring in Turbulent Business Environment: A Qualitative Study on Finnish SMEs. In M. Vemić (Ed.), *Optimal Management Strategies in Small and Medium Enterprises* (pp. 226–248). Hershey, PA: IGI Global. doi:10.4018/978-1-5225-1949-2.ch011

Tongiani, M. G., & Zhao, S. (2017). Traditional Italian Food Products on the Chinese Market: Best Practices for Italian Small and Medium Enterprises. In M. Vemić (Ed.), *Optimal Management Strategies in Small and Medium Enterprises* (pp. 22–43). Hershey, PA: IGI Global. doi:10.4018/978-1-5225-1949-2.ch002

Torlak, N. G. (2016). Improving the Role of Organisational Culture in Change Management through a Systems Approach. In A. Goksoy (Ed.), *Organizational Change Management Strategies in Modern Business* (pp. 230–271). Hershey, PA: IGI Global. doi:10.4018/978-1-4666-9533-7.ch012

Tran, B. (2016). Gendered Social-Networking Organizations: A View of the Sexed Mentorship Relationships. *International Journal of Organizational and Collective Intelligence*, *6*(2), 26–49. doi:10.4018/IJOCI.2016040103

Uyen, N. T., & Zainal, S. R. (2017). Non-State SMEs in Vietnam: Understanding Strategic Situation and Implications for Organizational Performance. In N. Ahmad, T. Ramayah, H. Halim, & S. Rahman (Eds.), *Handbook of Research on Small and Medium Enterprises in Developing Countries* (pp. 297–321). Hershey, PA: IGI Global. doi:10.4018/978-1-5225-2165-5.ch014

Vargas-Hernández, J. G. (2016). Critical Analysis of the Influence of Transnational Capitalism on Institutions and Organizations. In R. Das (Ed.), *Handbook of Research on Global Indicators of Economic and Political Convergence* (pp. 237–253). Hershey, PA: IGI Global. doi:10.4018/978-1-5225-0215-9.ch011

Vargas-Hernández, J. G. (2016). Strategic-Spatial Analysis of the Implementation of Business Opening Politics of Mexico. In R. Das (Ed.), *Handbook of Research on Global Indicators of Economic and Political Convergence* (pp. 274–313). Hershey, PA: IGI Global. doi:10.4018/978-1-5225-0215-9.ch013

Vargas-Hernández, J. G., Muratalla-Bautista, G., & Cruz, I. D. (2017). Point of View of Economical Organization at Manzanillo's Harbor. In R. Das (Ed.), *Social, Health, and Environmental Infrastructures for Economic Growth* (pp. 307–324). Hershey, PA: IGI Global. doi:10.4018/978-1-5225-2364-2.ch015

Vemić, M. B. (2017). A Further Look at Working Capital Optimization in Medium-Sized Firms: Concepts and Evidence. In M. Vemić (Ed.), *Optimal Management Strategies in Small and Medium Enterprises* (pp. 144–177). Hershey, PA: IGI Global. doi:10.4018/978-1-5225-1949-2.ch008

Vemić, M. B. (2017). Financial Innovation in Medium-Sized Enterprises Optimizes Their Gravitation Towards Capital Markets: Financial Future in Perspective. In M. Vemić (Ed.), *Optimal Management Strategies in Small and Medium Enterprises* (pp. 198–224). Hershey, PA: IGI Global. doi:10.4018/978-1-5225-1949-2.ch010

Villegas, M., & McGivern, M. H. (2015). Codes of Ethics, Ethical Behavior, and Organizational Culture from the Managerial Approach: A Case Study in the Colombian Banking Industry. *International Journal of Strategic Information Technology and Applications*, 6(1), 42–56. doi:10.4018/IJSITA.2015010104

Virkar, S. (2016). Can Codes of Ethical Conduct Work?: Evaluating the Effectiveness of Privatised Corporate Governance in a World of Political and Economic Convergence. In R. Das (Ed.), *Handbook of Research on Global Indicators of Economic and Political Convergence* (pp. 206–236). Hershey, PA: IGI Global. doi:10.4018/978-1-5225-0215-9.ch010

Windsor, D. (2017). Defining Corporate Social Responsibility for Developing and Developed Countries: Comparing Proposed Approaches. In D. Jamali (Ed.), *Comparative Perspectives on Global Corporate Social Responsibility* (pp. 1–27). Hershey, PA: IGI Global. doi:10.4018/978-1-5225-0720-8.ch001

Wolf, R., & Thiel, M. (2017). CSR in China: The Road to New Sustainable Business Models. In D. Jamali (Ed.), *Comparative Perspectives on Global Corporate Social Responsibility* (pp. 258–275). Hershey, PA: IGI Global. doi:10.4018/978-1-5225-0720-8.ch013

Wronka-Pośpiech, M. (2017). Applying Business Solutions to Social Problems: Social Co-Operative and Its Business Model – Evidence from Poland. In M. Lewandowski & B. Kożuch (Eds.), *Public Sector Entrepreneurship and the Integration of Innovative Business Models* (pp. 139–164). Hershey, PA: IGI Global. doi:10.4018/978-1-5225-2215-7.ch006

Yap, N. T., & Ground, K. E. (2017). Socially Responsible Mining Corporations: Before (or in Addition to) Doing Good, Do No Harm. In D. Jamali (Ed.), *Comparative Perspectives on Global Corporate Social Responsibility* (pp. 185–207). Hershey, PA: IGI Global. doi:10.4018/978-1-5225-0720-8.ch009

Yordanova, S. D. (2017). Optimizing Virtual Communities in Tourism to Facilitate Development of Small and Medium-Sized Enterprises. In M. Vemić (Ed.), *Optimal Management Strategies in Small and Medium Enterprises* (pp. 60–78). Hershey, PA: IGI Global. doi:10.4018/978-1-5225-1949-2.ch004

Yusoff, A., Ahmad, N. H., & Halim, H. A. (2017). Agropreneurship among Gen Y in Malaysia: The Role of Academic Institutions. In N. Ahmad, T. Ramayah, H. Halim, & S. Rahman (Eds.), *Handbook of Research on Small and Medium Enterprises in Developing Countries* (pp. 23–47). Hershey, PA: IGI Global. doi:10.4018/978-1-5225-2165-5.ch002

Zel, U. (2016). Leadership in Change Management. In A. Goksoy (Ed.), *Organizational Change Management Strategies in Modern Business* (pp. 272–288). Hershey, PA: IGI Global. doi:10.4018/978-1-4666-9533-7.ch013

About the Authors

Jerzy Kisielnicki is a professor of management has been the head of the Department of Information Systems in Management and Faculty of Management at Warsaw University. In 2002 postgraduate studies. Executive Education at Hardward Business School. His interests are organization and management, systems analysis, management information systems, process innovation (re-engineering), strategic management, and transition systems organization and management in market economy. Is a member of the Board of Organization and Management in Polish Academy of Science and is the head of the Scientific Council of Polish Society of Systems Information. He was involved and led many ICT projects. He has had about 220 publications. The most important publications in the last years: Decision management systems (ed. with J. Turin) Difin, 2012, Project management research and development, ed. Wolters Kluwer, 2013. Management ed. PWE, 2014.,, Management and Information ed. Placet 2015, Project management ed. Nieoczywista in 2017 Member of the editorial committees of the following international journals: International Journal of Electronic Business Publisher Intrcience Enterprises Ltd, UK, a member of the editorial committee of Information Science Reference in Hershey and New York.

Olga I. Sobolewska is an IT engineer and PhD in economics in the field of management science. A graduate of the Faculty of Management at the University of Warsaw. From 2014, she works at the Faculty of Management at the Warsaw University of Technology. A member of TNOiK (Scientific Society of Organization and Management) and NTIE (Scientific Society of Economic Information Technology). Her scientific specialty is the application of information systems in organizations, remote education and its use for the development of employees. She is interested in project management and organization's evolution towards a network of connections. She is particularly interested in the phenomena of cooperation and cooperation between organizations, the evolution of traditional organizations from various areas of the economy: business, science and public. She runs the Smart City workshop at the Warsaw University of Technique in Management Faculty. She is an author of publications in the field of project management and cooperation networks.

Index

R

relations 4, 10-11, 16, 27, 36, 45, 89-91, 94, 100, 119, 126, 147, 152-153

relationships 2-9, 12, 14-16, 30, 34, 49, 55, 88, 90, 94, 140, 147

report 11, 17, 60, 72-73, 79, 93, 95, 100, 145

representatives 12, 15, 18-19, 21, 41, 50, 74, 95

requirements 7, 38, 41, 45, 50, 58, 65, 104, 135, 147, 154-155

resources 5, 8-9, 12, 14, 16-17, 31-35, 40, 47, 49-50, 56-57, 59-60, 62-63, 65, 68, 76, 79, 86, 90, 96, 102, 104, 116-117, 122, 127, 134, 142, 146-147, 150

S

science 10-11, 15, 17-19, 21-22, 39, 41, 43, 45, 50, 87, 98, 117, 142-143, 150

solutions 3, 8, 11, 15-16, 26, 31-32, 34-36, 42, 57, 59, 61, 67-68, 73-76, 78-80, 88, 90, 96-100, 102, 115-119, 121-122, 127-128, 130-132, 134-135, 142, 155

structure 2-3, 5, 8-10, 14-17, 19, 21, 36, 39, 43, 55, 60-61, 65, 90-91, 118, 124, 129-131, 139-140, 142, 144-145, 152-155

structures 1, 3, 5, 7-9, 16-17, 19-20, 22, 43-44, 49, 55, 60, 80, 91, 97-98, 118, 129-130, 139-142, 144, 147, 152-153, 155

support 3, 5, 7, 57, 59, 73-76, 94, 99-100, 105, 109, 115, 118-119, 121-126, 130, 133-134, 147, 152, 154

T

technology 15, 21, 31, 36, 39, 42, 44, 60-61, 66, 68, 72-74, 80, 86, 99, 102, 115, 117, 122, 124, 139-142, 154

tools 21, 36, 39, 50, 57, 73-78, 88, 93, 96-97, 99, 115, 117-119, 124, 131, 133-134, 152

traditional 1, 3, 5, 7, 16, 35, 48-49, 55, 60, 86, 91, 98-99, 105, 107-108, 115-119, 126, 129-130, 140-141, 144, 153

transmission 69, 88, 90-91, 93, 96-97, 102, 107-109, 141

U

universities 8, 10, 12, 17, 38-39, 43, 50

W

Web 63, 65, 71, 73-76, 79

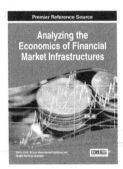

Stay Current on the Latest Emerging Research Developments

Become an IGI Global Reviewer for Authored Book Projects

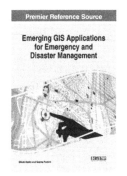

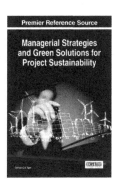

The overall success of an authored book project is dependent on quality and timely reviews.

In this competitive age of scholarly publishing, constructive and timely feedback significantly decreases the turnaround time of manuscripts from submission to acceptance, allowing the publication and discovery of progressive research at a much more expeditious rate. Several IGI Global authored book projects are currently seeking highly qualified experts in the field to fill vacancies on their respective editorial review boards:

Applications may be sent to:
development@igi-global.com

Applicants must have a doctorate (or an equivalent degree) as well as publishing and reviewing experience. Reviewers are asked to write reviews in a timely, collegial, and constructive manner. All reviewers will begin their role on an ad-hoc basis for a period of one year, and upon successful completion of this term can be considered for full editorial review board status, with the potential for a subsequent promotion to Associate Editor.

If you have a colleague that may be interested in this opportunity, we encourage you to share this information with them.

The Premier Reference for Information Science & Information Technology

Encyclopedia of
**Information Science
and Technology**
Fourth Edition

100% Original Content
Contains 705 new, peer-reviewed articles with color figures covering over 80 categories in 11 subject areas

Diverse Contributions
More than 1,100 experts from 74 unique countries contributed their specialized knowledge

Easy Navigation
Includes two tables of content and a comprehensive index in each volume for the user's convenience

Highly-Cited
Embraces a complete list of references and additional reading sections to allow for further research

Included in:

InfoSci®-Books

Encyclopedia of Information Science and Technology Fourth Edition

A Comprehensive 10-Volume Set

Mehdi Khosrow-Pour, D.B.A. (Information Resources Management Association, USA)
ISBN: 978-1-5225-2255-3; © 2018; Pg: 8,104; Release Date: July 31, 2017

The **Encyclopedia of Information Science and Technology, Fourth Edition** is a 10-volume set which includes 705 original and previously unpublished research articles covering a full range of perspectives, applications, and techniques contributed by thousands of experts and researchers from around the globe. This authoritative encyclopedia is an all-encompassing, well-established reference source that is ideally designed to disseminate the most forward-thinking and diverse research findings. With critical perspectives on the impact of information science management and new technologies in modern settings, including but not limited to computer science, education, healthcare, government, engineering, business, and natural and physical sciences, it is a pivotal and relevant source of knowledge that will benefit every professional within the field of information science and technology and is an invaluable addition to every academic and corporate library.

Scan for
Online Bookstore

Pricing Information

Hardcover: **$5,695**   E-Book: **$5,695**   Hardcover + E-Book: **$6,895**